Analog Circuits and Signal Processing

The Analog Circuits and Signal Processing book series, formerly known as the Kluwer International Series in Engineering and Computer Science, is a high level academic and professional series publishing research on the design and applications of analog integrated circuits and signal processing circuits and systems. Typically per year we publish between 5–15 research monographs, professional books, handbooks, edited volumes and textbooks with worldwide distribution to engineers, researchers, educators, and libraries.

The book series promotes and expedites the dissemination of new research results and tutorial views in the analog field. There is an exciting and large volume of research activity in the field worldwide. Researchers are striving to bridge the gap between classical analog work and recent advances in very large scale integration (VLSI) technologies with improved analog capabilities. Analog VLSI has been recognized as a major technology for future information processing. Analog work is showing signs of dramatic changes with emphasis on interdisciplinary research efforts combining device/circuit/technology issues. Consequently, new design concepts, strategies and design tools are being unveiled.

Topics of interest include:

Analog Interface Circuits and Systems;

Data converters;

Active-RC, switched-capacitor and continuous-time integrated filters;

Mixed analog/digital VLSI;

Simulation and modeling, mixed-mode simulation;

Analog nonlinear and computational circuits and signal processing;

Analog Artificial Neural Networks/Artificial Intelligence;

Current-mode Signal Processing;

Computer-Aided Design (CAD) tools;

Analog Design in emerging technologies (Scalable CMOS, BiCMOS, GaAs, heterojunction and floating gate technologies, etc.);

Analog Design for Test;

Integrated sensors and actuators;

Analog Design Automation/Knowledge-based Systems;

Analog VLSI cell libraries;

Analog product development;

RF Front ends, Wireless communications and Microwave Circuits;

Analog behavioral modeling, Analog HDL.

More information about this series at http://www.springer.com/series/7381

Leila Safari • Giuseppe Ferri • Shahram Minaei
Vincenzo Stornelli

Current-Mode Instrumentation Amplifiers

Springer

Leila Safari
Tehran, Iran

Shahram Minaei
Doğuş University
Istanbul, Turkey

Giuseppe Ferri
University of L'Aquila
L'aquila, Italy

Vincenzo Stornelli
University of L'Aquila
L'aquila, Italy

ISSN 1872-082X ISSN 2197-1854 (electronic)
Analog Circuits and Signal Processing
ISBN 978-3-030-01342-4 ISBN 978-3-030-01343-1 (eBook)
https://doi.org/10.1007/978-3-030-01343-1

Library of Congress Control Number: 2018957333

This Springer imprint is published by the registered company Springer Nature Switzerland AG
The registered company address is: Gewerbestrasse 11, 6330 Cham, Switzerland

Preface

Instrumentation amplifiers (IAs) are among the oldest and most widely used circuits that ensure a weak differential signal amplification in the presence of strong noises and common-mode signals. In the past, IAs were implemented using Op-Amps and resistors. Their major problem is the strict matching requirement between resistors to achieve a high common-mode rejection ratio (CMRR). Another limitation associated with Op-Amp-based IAs is their gain-dependent bandwidth. These weaknesses along with the rapid downscale of CMOS technology with reduced allowed supply voltages have made the Op-Amp-based IAs less attractive. Fortunately, after the emergence of current-mode signal processing, designers took the advantages of current-mode technique to mitigate the problems associated with Op-Amp-based IAs. Compared to conventional voltage-mode signal processing, the new technique showed interesting features such as wide frequency performance, simpler circuitry, low-voltage operation. This new generation of IAs based on current-mode signal processing is known as current-mode instrumentation amplifiers (CMIAs). Although signal processing is performed in current domain, the input and output signals in CMIAs can be current or voltage signals. Therefore, while benefiting from inherent advantages of current-mode signal processing such as low-voltage operation, high-frequency performance, simpler circuitry, CMIAs can cover a wide range of applications. The first CMIA was reported in 1989 by C. Toumazou and F. J. Lidgey, based on supply current sensing technique. This new CMIA utilized two Op-Amps and one resistor. It showed interesting features such as high CMRR without requiring tightly matched resistors and wide bandwidth independent of gain. Later, the famous Wheatstone bridge was also employed and modified to take advantage of current-mode signal processing. The current-mode Wheatstone bridge (CMWB) and mixed-mode Wheatstone bridge were introduced, being capable of operating with current-mode readout circuits. These advances provided the opportunity for nearly all types of sensor readout circuits to benefit from current-mode signal processing. Then, the CMIAs emerged as a rapidly advancing subject, so numerous topologies are found in literature, and new current-mode building blocks were introduced intended for CMIA applications. However, a book entirely and exclusively dedicated to the design of CMIAs is lacking. We decided to write this book to address the need for a guide on CMIA design and applications. It grew out employing the content of published journal and conference papers in the CMIA subject written by researchers all around the world. Our aim is to give an overall knowledge on CMIA design and make the comparison between various structures easier. The operation principle, advantages, and disadvantages of each topology are highlighted. Also, we have classified the reported CMIAs in four categories depending on their input and output signals. This classification simplifies selecting a specific topology for the desired application.

This book is written in nine chapters.

In the first chapter, the general definitions of the common-mode rejection ratio concept in single-ended, fully differential, and cascaded structures are studied. Then, we discuss about the limitations of the conventional Op-Amp-based IAs. The general classification of CMIAs is also developed in this chapter.

In Chap. 2, we cover supply current sensing technique which is the oldest and powerful method used in the design of CMIA. The basic concept and performance analysis of circuits based on this technique are discussed. The first and second generations of CMIAs developed using this technique are reviewed. Then, performance analysis of each generation as well as their limitations is studied. This chapter ends with an interesting comparison between the two generations.

In Chap. 3, we discuss about the Wheatstone bridges. First the fundamentals of conventional voltage-mode Wheatstone bridge (VMWB) and its readout circuits are studied. Then current-mode Wheatstone bridge (CMWB) principle, its readout circuit, and linearization technique are discussed. At the end of this chapter, the mixed-mode Wheatstone bridge principle and readout circuits which are aimed to take the benefits of both voltage-mode and current-mode signal processing are covered.

In Chap. 4, we consider the CMIA topologies designed with current conveyors. As the second-generation current conveyor (CCII) is considered the most famous and widely used current-mode building block, a chapter is solely dedicated to the CCII-based CMIA topologies. These CMIAs include topologies proposed by Wilson, Gift, Khan et al., Galanis and Haritantis, Su and Lidgey, Gkotsis et al., Koli and Halonen, etc. The principles and the effect of CCII's non-idealities on the overall CMRR of each topology are also studied.

In Chap. 5, all CMIA topologies based on various current-mode building blocks are described. These building blocks include operational floating current conveyor (OFCC), current differencing buffered amplifier (CDBA), current feedback operational amplifier (CFOA), operational transresistance amplifier (OTRA), differential difference current conveyor (DDCC), and differential voltage current conveyor (DVCC). The CMIAs studied in this chapter are classified according to their input and output signals.

In Chap. 6, we study CMIA structures with electronically tunable gain feature. Different methods used to achieve electronically tunable gain are discussed. The electronically tunable CMIAs are divided into two groups. In first group, gain is varied by the active building block, while, in second group, electronically variable resistors are used for this purpose.

In Chap. 7, we briefly discuss about the implications (and limitations) caused by mismatches in the CMIAs. Helpful techniques used to reduce the occurrence of mismatches are also reviewed. The chapter is finished by studying a CMIA topology with robust performance against mismatches.

In Chap. 8, we focus on the CMIAs designed for biomedical and low-voltage low-power applications. In particular, design considerations and challenges for biomedical applications are discussed. We study the realization of well-known bootstrapping technique in current-mode domain. Techniques used to improve noise performance of CMIAs are also reviewed. Then, various methods and implementations used to design low-voltage low-power CMIAs are studied.

In the final chapter, the CMIAs intended for sensor applications are reviewed. These include piezo-resistive, differential capacitive, ISFET, pH, and temperature sensors. For each application, background knowledge on the related sensor is given, and then the topology of the used CMIA is studied.

The intended audience for this book includes design engineers, researchers, and students. We tried our best to present each chapter in the simplest form as possible and independent of other chapters, so the readers are able to select and read a single chapter separately, based on their requirement.

We hope that this book will be helpful in inspiring new ideas in CMIA design and applications.

Tehran, Iran — Leila Safari
L'aquila, Italy — Giuseppe Ferri
Istanbul, Turkey — Shahram Minaei
L'aquila, Italy — Vincenzo Stornelli

Contents

Abbreviations

A_0	Open-loop gain
A/D	Analog to digital converter
CMRR	Common-mode rejection ratio
CMIA	Current-mode instrumentation amplifier
CMWB	Current-mode Wheatstone bridge
CCII	Second-generation current conveyor
CDBA	Current differencing buffered amplifier
CFOA	Current feedback operational amplifier
CMOS	Complementary metal oxide semiconductor
COA	Current operational amplifier
CDTA	Current differencing transconductance amplifier
CCCII	Current controlled current conveyor
CCDVCC	Current controlled differential voltage current conveyor
CCCCTA	Current controlled current conveyor transconductance amplifier
BiCMOS	Bipolar complementary metal oxide semiconductor
CDTRA	Current differencing transresistance amplifier
CFDITA	Current follower differential input transconductance amplifier
CMFB	Common-mode feedback
DDCC	Differential difference current conveyor
DVCC	Differential voltage current conveyor
ECG	Electrocardiogram
EEG	Electroencephalogram
ECCII	Electronically current gain controlled second-generation current conveyor
EX-CCCII	Extra-X current controlled current conveyor
FVF	Flipped voltage follower
FF	Flip-Flop
GBW	Gain-bandwidth product
HPF	High-pass filter
IA	Instrumentation amplifier
CNIC	Current negative impedance converter

I-I	Current input-current output
I-V	Current input-voltage output
ISFET	Ion-sensitive field-effect transistor
KVL	Kirchhoff's voltage law
KCL	Kirchhoff's current law
LPF	Low-pass filter
MZC-CDTA	Modified Z copy current differencing transconductance amplifier
MOS	Metal oxide semiconductor
PMOS	P-type metal oxide semiconductor
NFET	Natural FET
OFA	Operational floating amplifier
OFCC	Operational floating current conveyor
Op-Amp	Operational amplifier
OTRA	Operational transresistance amplifier
OC	Operational conveyor
NMOS	N-type metal oxide semiconductor
PSRR	Power supply rejection ratio
VMWB	Voltage-mode Wheatstone bridge
VCII	Second-generation voltage conveyor
V-V	Voltage input-voltage output
V-I	Voltage input-current output
RFET	Reference FET
SI-MO	Single input-multiple output

List of Figures

List of Tables

Chapter 1
Principles of Instrumentation Amplifiers

1.1 Principles and Applications of Instrumentation Amplifiers

IAs are used in many industrial and medical applications to measure and amplify small differential signals under the presence of large common-mode signals. The importance of an IA is that it can selectively amplify differential-mode signals and attenuate the unwanted common-mode noises and disturbances. It is to notice that, for example, the differential-mode output of devices like sensors and transducers is very low, while the common-mode signals usually have larger values. Therefore, without IA, the differential-mode signal will remain buried under larger common-mode signals. In the conventional IAs, the input and output signals are restricted to voltage type. However, the birth of current-mode signal processing has provided the possibility of designing current input/current output IAs. Here, the general specifications of IAs are summarized as follows:

1. **High CMRR:** Ideally an IA should have infinite CMRR. This indicates its capability to amplify only differential-mode signals and reject common-mode ones.
2. **Low-Voltage Low-Power Consumption:** Reduced allowed supply voltage in modern CMOS technologies along with relatively large threshold voltage of transistors, make the low voltage operation a necessity for IA circuit. In addition, low-voltage low-power operation are very important especially for portable applications to extend battery life.
3. **High PSRR:** This feature indicates the capability of an IA to reject noises and disturbance on supply rails.
4. **High and Variable Differential-Mode Gain:** A good IA should provide high differential-mode gain which can be easily varied.

G. Ferri et al., *Current-Mode Instrumentation Amplifiers*, Analog Circuits and Signal Processing, https://doi.org/10.1007/978-3-030-01343-1_1

5. **Low Input Referred Noise and Offset:** Due to the very low value of differential-mode signals to be measured, the input referred noise and offset of an IA should be very low.
6. **High Input Impedance:** To avoid loading effect of IA on the signal to be amplified (for example, coming from the sensors), a very high input impedance is required for voltage input IA. For voltage inputs, ideally the input impedance should be infinity.
7. **Low Output Impedance:** For voltage output IAs, the output impedance should be very low (ideally zero) to reduce loading effect on the next stage.

Figure 1.1 shows a summary of applications benefiting from IAs. They appear in wide range of industrial and medical applications as stated in [1]. A common industrial application for IAs is data acquisition. IAs are used, for example, in amplifying low level output signals of pressure (resistive) or temperature sensors connected in a Wheatstone Bridge configuration. An example of medical application for IAs is in electrocardiography machines, or ECGs. Other application examples include Monitor and Control Electronics, Software Programmable Applications, Audio Applications, High Speed Signal Conditioning, Video Applications, Power Control Applications [1].

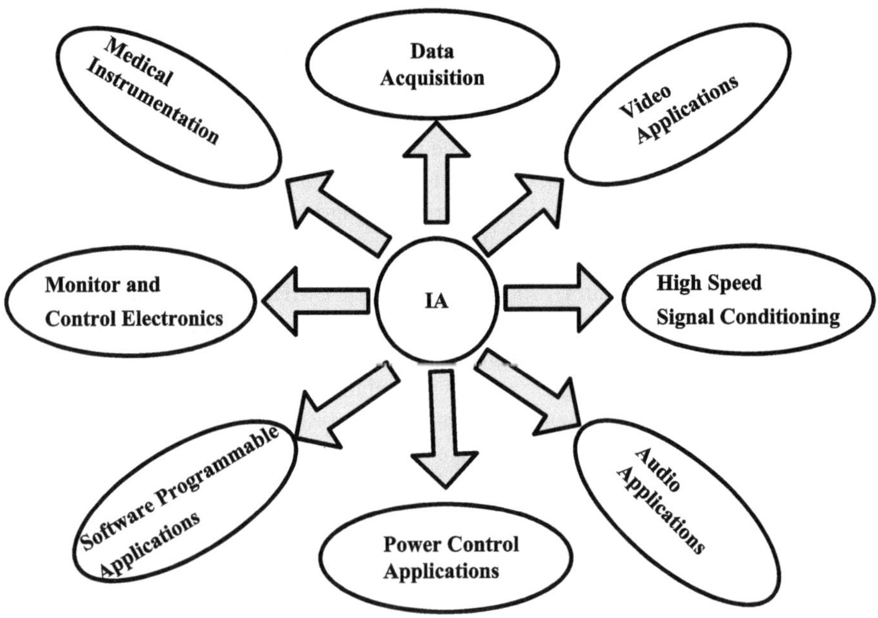

Fig. 1.1 Applications of IAs

1.2 Common-Mode Rejection Ratio

1.2.1 Definitions of Common-Mode Rejection Ratio

1.2.1.1 Single Ended Amplifiers

Common-Mode Rejection Ratio is one of the most important specifications of a differential amplifier. For a single ended amplifier, CMRR indicates the amplifier capability to suppress undesirable common-mode signals and to amplify differential-mode ones. Figure 1.2 shows the general schematic of a differential amplifier with single ended output. Based on the application and type of the IA (Voltage-Mode or Current-Mode), the input and output signals can be of either voltage or current kind. For voltage-input voltage-output IA, the signal named V_{id} is the differential input voltage which is selectively amplified with gain of A_d to produce output signal V_{od} as follows:

$$A_d = \frac{V_{od}}{V_{id}} \tag{1.1}$$

The unwanted common-mode voltage signals and disturbances represented by V_{ic} are presented equally at both inputs. These signals are attenuated by common-mode gain A_C to produce the output voltage signal V_{oc}. A_C is typically much smaller than A_d and is expressed as:

$$A_C = \frac{V_{oc}}{V_{ic}} \tag{1.2}$$

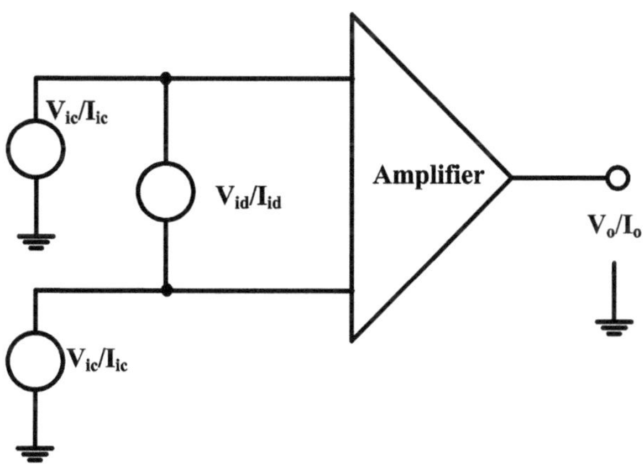

Fig. 1.2 General Schematic of a single output differential amplifier where the signal can be of voltage/current kind

CMRR is defined as the ratio of A_d to A_c and is expressed in dB as:

$$CMRR = 20\log\left(\frac{A_d}{A_c}\right) \tag{1.3}$$

The definition of CMRR for current-input and/or current-output IAs can be simply obtained by replacing voltage signals with current signals in Eq. (1.1) and Eq. (1.2).

1.2.1.2 Fully Differential Amplifiers

The general schematic of a fully differential amplifier is shown in Fig. 1.3 which is a four-terminal device. Its CMRR definition is given in [2], where for simplification, the inputs are considered as voltage signals. Similar definitions hold for current signals. According to [2], the input signals v_1, v_2 can be expressed in terms of their common-mode and differential-mode components as:

$$v_1 = v_c + \frac{v_d}{2} \tag{1.4}$$

$$v_2 = v_c - \frac{v_d}{2} \tag{1.5}$$

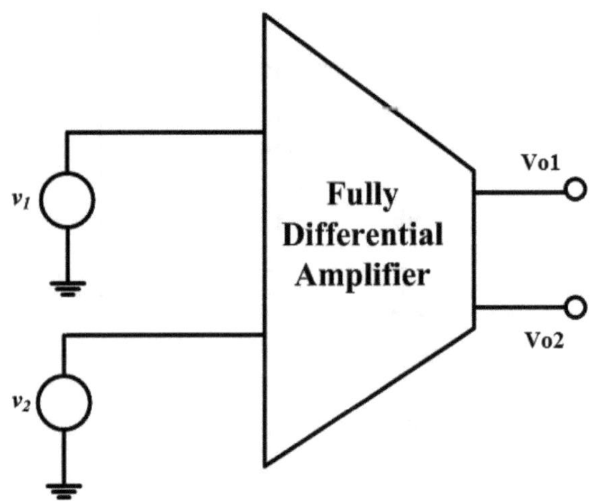

Fig. 1.3 General schematic of a fully differential amplifier

where $v_d = v_1 - v_2$ is the differential-mode and $v_c = (v_1 + v_2)/2$ is the common-mode part of the input signals. Differential output voltages are named V_{o1}, V_{o2}.The produced common-mode and differential-mode output voltages are:

$$V_d = A_{dc}v_c + A_{dd}v_d \tag{1.6}$$

$$V_c = A_{cc}v_c + A_{cd}v_d \tag{1.7}$$

where $V_d = V_{o1} - V_{o2}$ and $V_C = (V_{o1} + V_{o2})/2$ are the differential-mode and common-mode components of output signals, respectively. A_{dd} is the amplifier differential-mode gain. It is defined as the ratio of differential-mode output signal to the applied differential-mode input signal while the common-mode input signal is zero, while A_{cc} is the amplifier common-mode gain and is defined as the ratio of output common-mode signal to the input common-mode signal while no differential-mode input signal is applied. The parameter A_{cd} is the differential-mode to common-mode conversion gain and is expressed as the ratio of produced common-mode output signal to the applied differential-mode input. A_{dc} is the common-mode to differential-mode conversion gain. It can be achieved by applying common-mode input signal and measuring the produced differential-mode output signal. For a non-zero A_{dc}, a common-mode input signal produces a differential-mode signal at the output that is considered as an error signal. For those applications where the differential-mode output is of primary interest, the value of this error signal must be as low as possible. Therefore, for a fully differential amplifier, the definition of CMRR is [2]:

$$CMRR = \frac{A_{dd}}{A_{dc}} \tag{1.8}$$

Here, CMRR indicates how much of the undesired common-mode signal will appear as differential-mode output. Unlike the single-ended amplifier, in a fully differential amplifier the effects of A_{cc} and A_{cd} are not considered in CMRR definition. The reason is that the outputs produced by these gains are equally present at both outputs and therefore in the fully differential output structure their effect is canceled out.

1.2.2 Common-Mode Rejection Ratio for Cascaded Stages

In [3] a general rule to calculate the CMRR in cascaded stages is driven. A circuit with differential input and differential output is described by four parameters (G_{DD}, G_{CC}, G_{CD} and G_{DC}). G_{DD} is the differential-mode gain which is determined by the ratio of differential-mode output to differential-mode input when no common-mode input is applied. G_{CC} is the common-mode gain defined as the ratio of common-mode output to common-mode input when differential-mode input is zero. G_{DC} is the common-mode to differential gain which is the output differential-mode signal

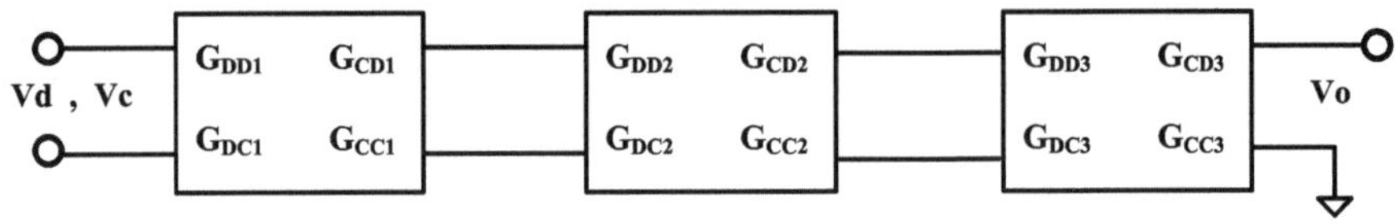

Fig. 1.4 Three stage amplifier [3]

due to common-mode input signal. Finally, G_{CD} is the differential to common-mode gain which is the ratio of common-mode output due to differential-mode input. It is shown in [3] that the total CMRR of an n-stage amplifier with single ended output is deducible from the following relation:

$$\frac{1}{CMRR_T} \approx \frac{1}{CMRR_1} + \frac{1}{CMRR_2} + \ldots + \frac{1}{CMRR_n} \tag{1.9}$$

being CMRR_n the common-mode rejection ratio of the nth stage expressed as:

$$CMRR_n = C_n \prod_{i=1}^{n-1} D_i \tag{1.10}$$

where:

$$C_n = \frac{G_{DDn}}{G_{DCn}}, \quad D_i = \frac{G_{DDi}}{G_{CCi}} \tag{1.11}$$

For example, the total CMRR of the three-stage amplifier shown in Fig. 1.4 is [2]:

$$\frac{1}{CMRR_T} \approx \left(\frac{G_{DD1}}{G_{DC1}}\right)^{-1} + \left(\frac{G_{DD1}}{G_{CC1}}\frac{G_{DD2}}{G_{DC2}}\right)^{-1} + \left(\frac{G_{DD1}}{G_{CC1}}\frac{G_{DD2}}{G_{CC2}}\frac{G_{DD3}}{G_{DC3}}\right)^{-1} \tag{1.12}$$

From Eq. (1.12) it is deduced that to increase the overall CMRR of a cascaded amplifier, the CMRR of the first stage is very critical. That is, to improve the overall CMRR, we should have $G_{DD1}/G_{DC1} \gg 1$. The same condition allows to reduce the effect of other stages on CMRR.

1.3 Conventional 3-Op-Amp Based Voltage-Mode Instrumentation Amplifier

1.3.1 CMRR Analysis

Figure 1.5 shows the popular 3-Op-Amp instrumentation amplifier. In [4] the CMRR of this circuit is determined by considering Op-Amps finite common-mode and differential-mode gains. The analysis is done by considering the whole circuit

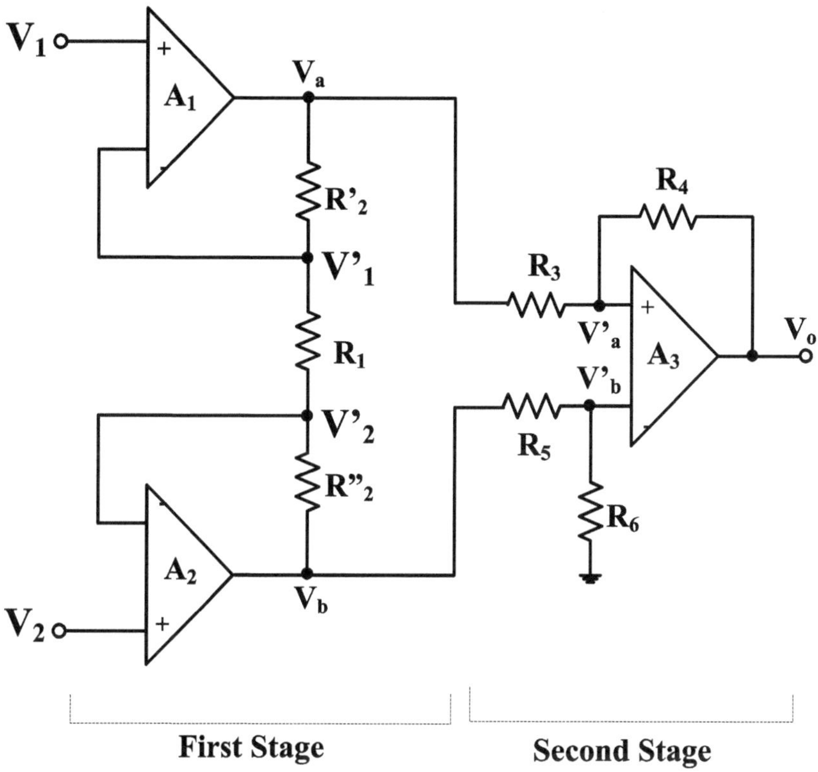

Fig. 1.5 Conventional 3 Op-Amp based voltage-mode instrumentation amplifier [4]

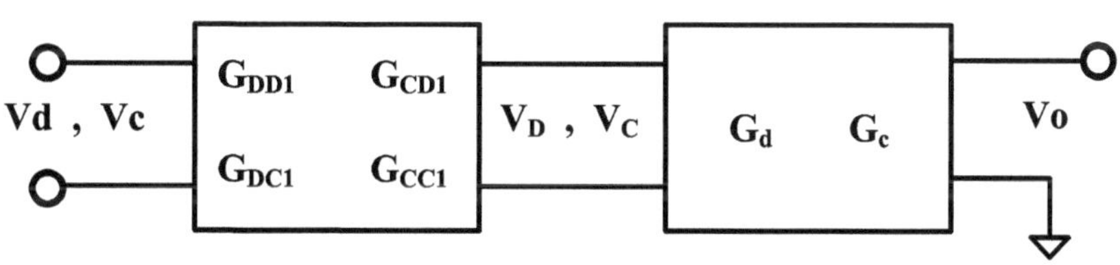

Fig. 1.6 Simplification of conventional 3 Op-Amp based IA as a two-stage amplifier [4]

as a cascade of one fully differential amplifier and a single ended amplifier as is shown in Fig. 1.6. According to the general rule of *CMRR* in cascaded amplifiers presented in Sect. 1.2.2, the total CMRR of the two-stage amplifier of Fig. 1.6 is related to that of the first ($CMRR_F$) and second ($CMRR_S$) stages as follows:

$$\frac{1}{CMRR_T} \approx \frac{1}{CMRR_F} + \frac{1}{CMRR_S} = \left(\frac{G_{DD1}}{G_{DC1}}\right)^{-1} + \left(\frac{G_{DD1}}{G_{CC1}}\frac{G_D}{G_C}\right)^{-1} \tag{1.13}$$

In [4], by writing the relation between voltages specified at different nodes of Fig. 1.5, the parameters G_{DD1}, G_{CD1}, G_{DC1} and G_{CC1} are derived for the first stage. Using these parameters and by the assumptions that $A_d \gg A_c$ and $CMRR_3 CMRR_R \gg 1/4$, the simplified overall CMRR of 3-Op-Amp based instrumentation amplifier is given, considering (1.13), as:

$$\frac{1}{CMRR_T} \approx \frac{1}{A_{d1}} - \frac{1}{A_{d2}} + \frac{1}{CMRR_1} - \frac{1}{CMRR_2} + \frac{1}{GCMRR_3} + \frac{1}{GCMRR_R} \tag{1.14}$$

where:

$$G \approx 1 + \frac{R_2'}{R_1} + \frac{R_2''}{R_1} \tag{1.15}$$

$$CMRR_R = \frac{1}{2}\frac{2R_4R_6 + R_4R_5 + R_3R_6}{R_3R_6 - R_4R_5} \tag{1.16}$$

being $CMRR_R$ the CMRR value of the resistors and $CMRR_i$ that of Op-Amp A_i (i = 1,2,3) defined as:

$$CMRR_i = \frac{A_{di}}{A_{ci}} \tag{1.17}$$

In Eq. (1.14), the first four terms are related to first stage and the last two are related to second stage. According to Eq. (1.14), the best approach to increase the overall CMRR is to use well matched Op-Amps in terms of differential and common-mode gains. In addition, the high values for $CMRR_3$ and $CMRR_R$ are helpful in increasing overall CMRR. It can be noticed that by increasing the value of G, the effects of $CMRR_R$ and $CMRR_3$ on total *CMRR* are reduced.

1.3.2 Differential Gain Analysis

In the classical 3-Op-Amps IA of Fig. 1.5, the input Op-Amps are configured in non-inverting configuration while the third Op-Amp operates as a difference amplifier. By assuming $R_3 = R_5$ and $R_4 = R_6$, V_{out} can be written as:

$$V_{out} = \left(V_a - V_b\right)\frac{R_4}{R_3} \tag{1.18}$$

V_a and V_b can be obtained by applying KVL and KCL analysis to input Op-Amps and using the well-known superposition theorem as follows. If $V_2 = 0$,

V'_2 is at virtual ground and A_1 operates as a non-inverting amplifier. Therefore, V_a is found as:

$$V_a = \left(1 + \frac{R_2'}{R_1}\right) V_1 \tag{1.19}$$

If $V_1 = 0$, and V_2 is applied, V'_1 is at virtual ground and V_a is obtained as:

$$V_a = -\left(\frac{R_2'}{R_1}\right) V_2 \tag{1.20}$$

Applying the superposition theorem, V_a and V_b are achieved as:

$$V_a = \left(1 + \frac{R_2'}{R_1}\right) V_1 - \left(\frac{R_2'}{R_1}\right) V_2 \tag{1.21}$$

$$V_b = \left(1 + \frac{R_2''}{R_1}\right) V_2 - \left(\frac{R_2''}{R_1}\right) V_1 \tag{1.22}$$

from which the V_{out} is found as:

$$V_{out} = \left(1 + \frac{R_2' + R_2''}{R_1}\right) \frac{R_4}{R_3} (V_1 - V_2) \tag{1.23}$$

For $V_1 = -V_2 = V_d/2$, the differential-mode gain is:

$$A_d = \frac{V_{out}}{V_d} = \left(1 + \frac{R_2' + R_2''}{R_1}\right) \frac{R_4}{R_3} \tag{1.24}$$

According to Eq. (1.24), resistor R_1 can be used to vary differential-mode gain.

1.3.3 Gain Bandwidth Conflict

The classical 3-Op-Amp based IA of Fig. 1.5 suffers from a fixed gain-bandwidth product. This means that its −3 dB bandwidth is proportionally reduced by increasing gain. This problem arises because input Op-Amps A_1-A_2 are in non-inverting feedback configuration which presents a constant gain bandwidth product based on the negative feedback theory as is explained in [5, 6]. Figure 1.7 shows the input inverting amplifiers for differential-mode inputs ($V_1 = -V_2 = V_d/2$). Using the

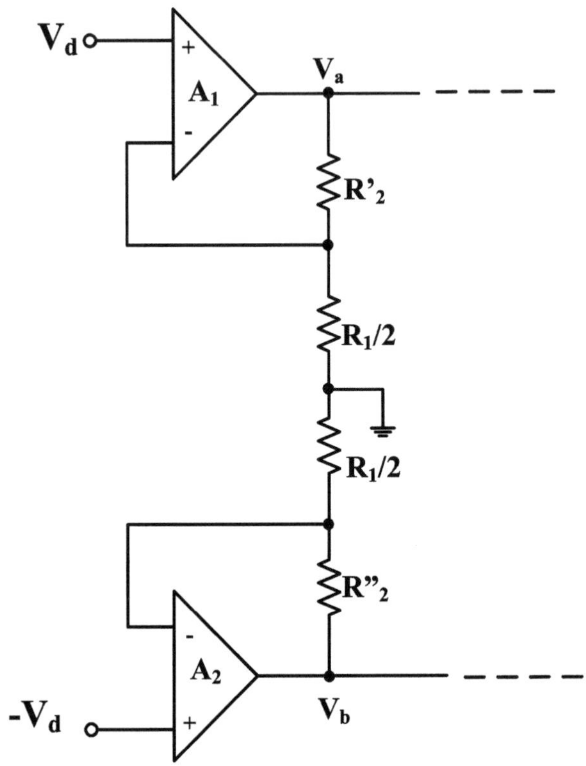

Fig. 1.7 Classical 3 Op-Amp IA for differential-mode inputs

general analysis of feedback circuits [6], the gain and −3 dB frequency of these inverting amplifiers are obtained respectively as:

$$G = 1 + \frac{R_2}{R_1/2} \tag{1.25}$$

$$\omega_{-3\text{dB}} = \frac{\omega_{GBW}}{1 + \frac{R_2}{R_1/2}} = \frac{\omega_{GBW}}{G} \tag{1.26}$$

Here $\boldsymbol{\omega}_{-3\text{dB}}$ is the bandwidth of non-inverting amplifier, $\boldsymbol{\omega}_{\text{GBW}}$ is the unity gain frequency of Op-Amps A_1-A_2 and $R'_2 = R''_2 = R_2$. The gain is usually varied by R_1. From Eqs. (1.25) and (1.26), we recognize the critical limitation of classic IA i.e. the reduction of −3 dB bandwidth by increasing gain.

1.4 Current-Mode Instrumentation Amplifiers (CMIAs)

1.4.1 Features

In recent years we have evidenced ever increasing prominence of current-mode signal processing due to its capability to operate at low supply voltages. In current-mode circuits the signals are represented by current and due to low impedance at internal nodes, voltage swings are minimized [7–9]. This makes current-mode circuits highly compatible with modern CMOS technologies where there is a strict limitation on the allowable supply voltages. In addition, many operations such as subtraction can be implemented in a much simpler way by current signals compared to voltage signals. This property makes the current-mode signal processing the best technique for IA design. The reason is that IAs attenuate the unwanted common-mode input signals by subtraction action. To clarify this matter, let us consider current subtraction circuit of Fig. 1.8 as an example realized by a simple current mirror. One of the inputs is applied to current mirror input while the other one is applied to its output. In fact the current mirror is used to produce the inversion of I_1 which is added to I_2 at output node producing $I_1 - I_2$ at output. For common-mode signals of $I_1 = I_2 = I_{cm}$, the output signal is zero. This technique can be seen in the CMIA reported in [10]. It is also possible to perform voltage subtraction very easily employing various current-mode active building blocks [11–15] as it will be discussed in next chapters.

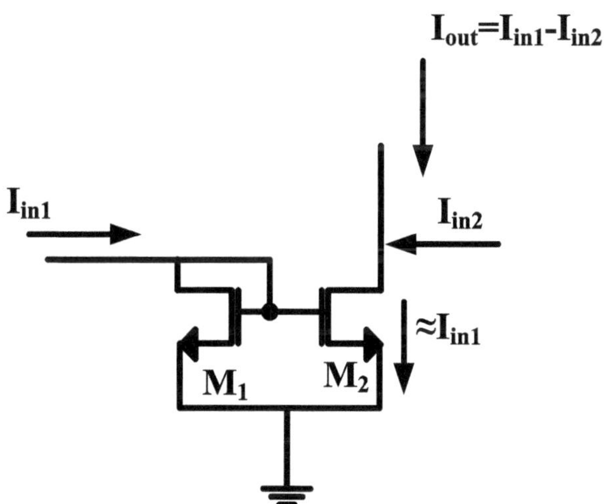

Fig. 1.8 A simple current subtractor realized by current mirror

The second advantage of the current-mode technique is that they do not require high impedance nodes and therefore their bandwidth has the potential for approaching maximum possible values. Therefore, IAs using current-mode signal processing can achieve high bandwidth. Most importantly, unlike conventional Op-Amp based IAs, in the CMIAs the bandwidth is independent of gain. This is due to the fact that in CMIAs the active building blocks are not in negative feedback configuration. Their capability to maintain constant bandwidth independent of gain makes CMIAs suitable for higher frequency applications.

1.4.2 Classification

Based on the input and output signals, the CMIAs can be categorized in four main types as follows:

1. Voltage Input – Voltage Output (V-V) CMIA: The input and output signals are voltage but the signal processing is performed in current domain. High input impedance (ideally infinite) and low output impedance (ideally zero) are required for V-V CMIA.
2. Current Input – Current Output (I-I) CMIA: The input and output signals are currents and the signal processing is performed in current domain, therefore it can be called "pure" CMIA. The low input (ideally zero) and high output (ideally infinite) impedances are required for this type.
3. Current Input – Voltage Output (I-V) CMIA: The input signal is current and the output signal is voltage. This type is called trans-impedance (or trans-resistance) CMIA. It has low input (ideally zero) and low output (ideally zero) impedances.
4. Voltage Input – Current Output (V-I) CMIA: The input signal is voltage and the output signal is current. This type is called trans-conductance CMIA. It has high input (ideally infinite) and high output (ideally infinite) impedances.

Each of these types can be designed to have electronically tunable gain where differential-mode gain can be varied by means of a control voltage or current. Chapter 6 is dedicated to various approaches to achieve this property.

References

1. Kitchin C., Counts L. (2006), A Designer's guide to instrumentation amplifiers, 3rd Edition, Analog Device, USA.
2. Schoenfeld R. L. (1970), Common-Mode Rejection ratio-two definitions, IEEE Transactions on Biomedical Engineering, BME-17(1):73–74.
3. Pallas-Areny R., Webster J. G. (1991), Common-Mode rejection ratio for cascaded differential amplifier stages, IEEE Transactions on Instrumentation and Measurement, 40(4):677–681.

4. Pallas-Areny R., Webster J. G. (1991),Common-Mode rejection ratio in differential amplifiers, IEEE Transactions on Instrumentation and Measurement, 40(4):669–676.
5. Pennisi S., Scotti G., Trifiletti A. (2011), Avoiding the Gain-Bandwidth trade off in feedback amplifiers, IEEE Transactions on Circuits and Systems I: Regular Papers, 58(9): 2108–2113.
6. Bruun E. (1993), Feedback analysis of transimpedance operational amplifier circuits, IEEE Transactions on Circuits and Systems I: Fundamental Theory and Applications, 40(4):275–278.
7. Wilson B., Lidgey F. J., Toumazou C. (1988), Current mode signal processing circuits, IEEE International Symposium on Circuits and Systems, Espoo, Finland, 3:2665–2668.
8. Arbel A. F. (1991), Current-mode signal processing, 17th Convention of Electrical and Electronics Engineers in Israel, Tel Aviv, 1991.
9. Toumazou C., Lidgey F. J., Haigh D. (1992), Analogue IC design: the current-mode approach, IET.
10. Safari L., Minaei S. (2013), A novel resistor-free electronically adjustable current-mode instrumentation amplifier, Circuits, Systems and Signal Processing, 32(3):1025–1038.
11. Ghallab Y. H., Badawy W., Kaler K. V. I. S., Maundy B. J. (2005), A novel current-mode instrumentation amplifier based on operational floating current conveyor,IEEE Transactions on Instrumentation and Measurement, 54(5):1941–1949.
12. Cini U. (2014), A low-offset high CMRR current-mode instrumentation amplifier using differential difference current conveyor, *IEEE International Conference on Electronics, Circuits and Systems (ICECS)*, 2014.
13. Hassan T., Mahmoud S. A. (2014), New CMOS DVCC realization and applications to instrumentation amplifier and active-RC filters, International Journal of Electronics and Communication (AEÜ), 64:47–55.
14. Yang T. Y., Liu K. Y., Wang H. Y. (2011), Novel high-CMRR DVCC-based instrumentation amplifier, 2nd International Conference on Engineering and Industries (ICEI), 2011.
15. Yuce E.(2011), Various current-mode and voltage-mode instrumentation amplifier topologies suitable for integration, Journal of Circuits, Systems, and Computers,19(3):689–699.

Chapter 2
CMIA Based on Op-Amp Power Supply Current Sensing Technique

2.1 Op-Amp Power Supply Current Sensing Technique

2.1.1 Basic Concept

In vast majority of operational amplifiers, output transistor current is not independently accessible. Op-Amp power supply current sensing is a simple technique for detecting output transistor current from the Op-Amp power supply leads. This technique was first introduced in 1979 to implement voltage controlled current sources and current controlled current sources [1]. It has been implemented by connecting current mirrors to Op-Amp supply leads and makes possible to gain access to output transistor current separately.

Figure 2.1 shows a typical circuit implementation of the supply current sensing technique, in which the Op-Amp operates as a voltage follower [2]. To simplify the diagram, current-mirror symbols are used. An examination of Fig. 2.1 indicates that when V_{in} is nonzero, a current I_{o1} flows in R_1. The produced current I_{o1} modifies the bias current at Op-Amps positive and negative supply leads. Under this condition, an imbalance in bias currents occurs as is shown in Fig. 2.1. By the assumption that the used current mirrors have unity gain, according to Kirchhoff current law, the output current I_{o2} is exactly equal to I_{o1}, for any R_1 and R_2. Therefore, in this way, load current can be retrieved from bias leads. In case when $V_{in} = 0$ so that $I_{o1} = 0$, the Op-Amp positive and negative supply bias currents will be equal in magnitude. Therefore, output branches of upper and lower current mirrors will have equal bias currents and the current in R_2 will be equal to zero.

The advantage of the circuit of Fig. 2.1 is that its bandwidth is maximum because of the employed full negative feedback. This technique has been shown by many authors to be extremely useful technique in realizing many high performance analogue building-blocks. For example, it has been used in [3–7] to design current conveyors, active filters, precision rectifiers and other analogue circuits. The current conveyors implemented using this technique exhibited superior performance

G. Ferri et al., *Current-Mode Instrumentation Amplifiers*, Analog Circuits and Signal Processing, https://doi.org/10.1007/978-3-030-01343-1_2

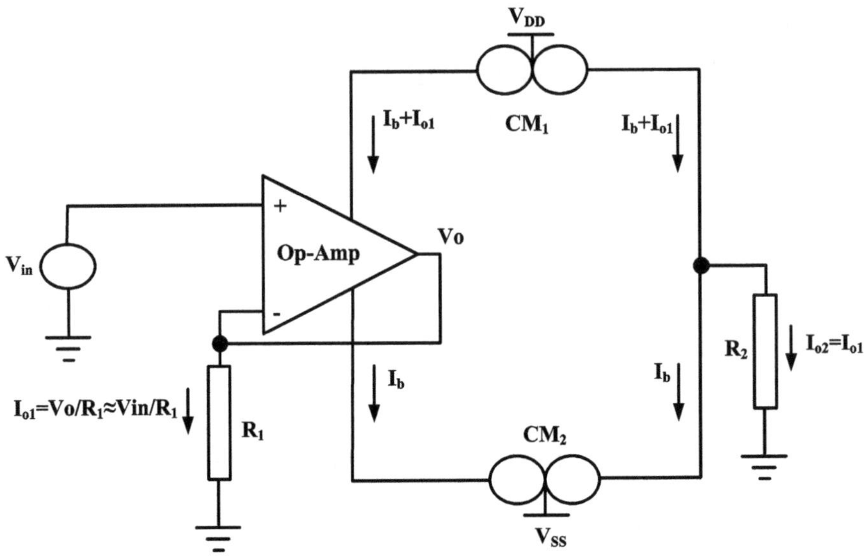

Fig. 2.1 Op-Amp power supply current sensing technique [2]

compared to earlier implementations of current conveyors which suffered from a complicated circuit, poor frequency performance and the requirement for tightly matched resistors [8].

2.1.2 Analysis of Op-Amp Power Supply Current Sensing Technique

The equivalent circuit of Fig. 2.1 is shown in Fig. 2.2 in which all non-idealities affecting the operation of the Op-Amp power supply current sensing technique are modeled. In the equivalent circuit of Fig. 2.2, the Op-Amp finite common-mode rejection ratio (CMRR), finite power-supply rejection ratio (PSRR) and non-zero input offset voltage (*Vos*) are extracted to input terminal and current mirrors CM_1 and CM_2 are shown as dependent current sources with finite output impedances (r_{o1} and r_{o2}) and non-unity gains (K_1 and K_2), respectively. The Op-Amps input bias current is assumed to be zero (as is always in the case of implementation with CMOS technology).

In order to model the effect of finite CMRR, we consider the common-mode output voltage V_{oc} as:

$$V_{oc} = V_{inc} A_c \tag{2.1}$$

In Eq. (2.1), A_c is the Op-Amps common-mode gain and V_{inc} is the common-mode voltage at the input terminals of Op-Amp which is equal to:

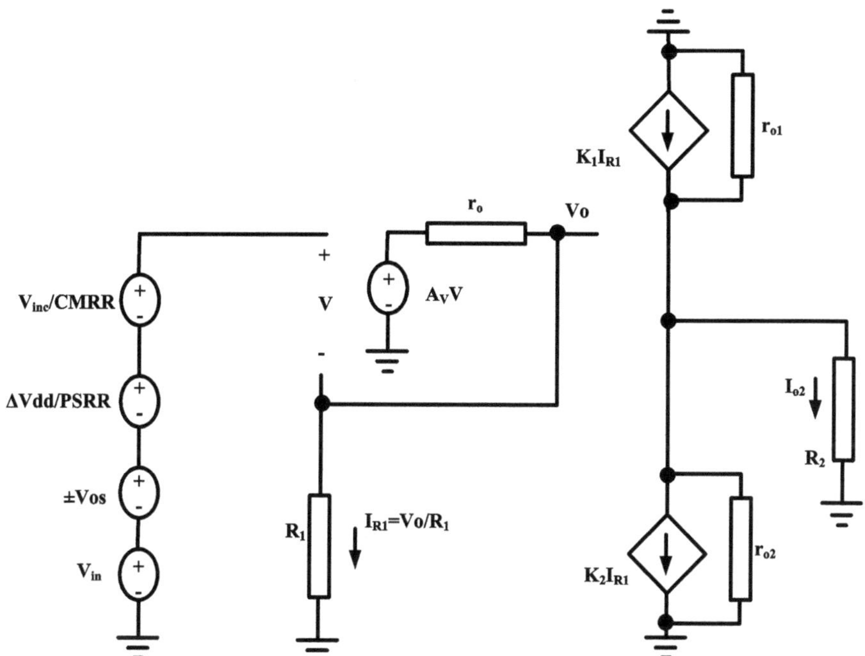

Fig. 2.2 Equivalent circuit of Fig. 2.1

$$V_{inc} = \frac{V_{in} + V_o}{2} \tag{2.2}$$

By considering the definition of CMRR:

$$CMRR = \frac{A_d}{A_c} \tag{2.3}$$

and inserting A_c from Eq. (2.3) into Eq. (2.1), V_{oc} is found as:

$$V_{oc} = \frac{V_{inc}}{CMRR} A_d \tag{2.4}$$

From Eq. (2.4), the effect of finite CMRR can be modeled as an input voltage equal to $V_{inc}/CMRR$ at Op-Amps input terminal. Similarly, the effect of Op-Amps finite PSRR can be modeled as an input voltage equal to $\Delta V_{dd}/PSRR$ in which ΔV_{dd} is the variation in supply voltage.

From Fig. 2.2, V_o can be written as:

$$V_o = \frac{R_1}{R_1 + r_o} A_V V \tag{2.5}$$

where V is equal to:

$$V = V_{in} \pm V_{os} + \frac{\Delta Vdd}{PSRR} + \frac{V_{inc}}{CMRR} - V_o \tag{2.6}$$

By inserting Eqs. (2.2) and (2.6) into Eq. (2.5), V_o can be found as:

$$V_o = V_{in} \frac{A_V \frac{R}{R+r_o}\left(1 + \frac{1}{2CMRR}\right)}{\left(1 - A_V \frac{R}{R+r_o}\frac{1}{2CMRR} + A_V \frac{R}{R+r_o}\right)} + \left(\frac{\Delta Vdd}{PSRR} \pm V_{os}\right)\frac{A_V \frac{R}{R+r_o}}{\left(1 - A_V \frac{R}{R+r_o}\frac{1}{2CMRR} + A_V \frac{R}{R+r_o}\right)} \tag{2.7}$$

As it is seen from Fig. 2.2, V_o is converted to current by R_1. The produced current in R_1 is conveyed to output node by current mirrors and we have:

$$I_{o2} = K_1\left(\frac{r_{o1}}{r_{o1} + R_2}\right)\left(\frac{V_o}{R_1}\right) \quad \text{for } V_{in} > 0 \tag{2.8}$$

$$I_{o2} = K_2\left(\frac{r_{o2}}{r_{o2} + R_2}\right)\left(\frac{V_o}{R_1}\right) \quad \text{for } V_{in} < 0 \tag{2.9}$$

As it can be deduced from Eqs. (2.8) and (2.9), the output current at R_2 experiences distortion mainly because it is very difficult, if not impossible, to match p-type and n-type current mirrors.

2.1.3 Limitation and Design Challenges

There are several problems with Op-Amp power supply current sensing technique shown in Fig. 2.1. The main one is due to its high supply voltage requirement and low dynamic range. The two current mirrors connected in series with Op-Amps supply leads increase the required supply voltage. Because of the reduced allowed supply voltage and transistors high threshold voltage in modern technologies [9, 10], employing this technique is very difficult. The other limitation is related to noise performance because the noise of entire circuit is transferred to the output. In addition, the current transfer accuracy I_{R2}/I_{R1} is not precise enough, because the finite output resistances of the two current mirrors affect the output current. Therefore, current mirrors must be designed to have very high output impedance: this is other major challenge, because the intrinsic output impedance of transistors in fine technologies is low [9]. Furthermore, to avoid distortion, there must be good matching between current mirrors which is not easy to maintain. In addition, the accuracy of the voltage transfer ratio V_o/V_{in} and non-zero currents at Op-Amps supply leads when R_1 is infinite also affects the current transfer accuracy. It is evident that, for infinite R_1, the output current must be zero. However, as it is explained in [11], due to the finite output impedance of current sources in the Op-Amps internal circuit, currents at supply leads is modulated by input signal and introduce error to the output current. To overcome this problem, common-mode bootstrapping technique can be applied to conventional Op-Amp in [12]. Significant improvement in current and voltage transfer accuracy is achieved at the expense of reduced common-mode voltage range.

Finally, according to Eq. (2.7), high voltage transfer accuracy requires a high-performance Op-Amp with very high (ideally infinite) CMRR, PSRR and differential gain which is quite challenging to design.

2.2 CMIA Based on Op-Amp Power Supply Current Sensing

2.2.1 Introduction

To describe the operation principle of CMIA based on supply current sensing technique let us consider the input stage of the conventional 3-Op-Amps IA shown in Fig. 2.3. The input Op-Amps operate as voltage buffers and transfer the input signals to nodes A and B [13]. For common-mode input of $V_1 = V_2 = V_{cm}$, the voltages at nodes A and B approximately will be $V_A = V_B = V_{cm}$. The second stage in the conventional three Op-Amp instrumentation amplifier subtracts V_A from V_B and produces a single-ended output proportional to $V_A - V_B$. The second stage is usually a standard four resistor differential amplifier. To have a very low common-mode gain at second stage, well-matched resistors are required. However, it is interesting to note that the current produced through R is simply given by:

$$I = \frac{V_1 - V_2}{R} \tag{2.10}$$

For $V_1 = V_2 = V_{cm}$, the current I will be zero. Therefore, if current I is considered as an output parameter, the circuit will have zero common-mode transconductance gain [13]. The current I can be accessed through Op-Amp power supply current sensing technique. For the first time in 1989, this method was used by C. Toumazou et al. to design a CMIA [13]. First generation of CMIA based on Op-Amp power supply current sensing technique has a single output topology [13–18] while second generation one has a balanced structure at the output [19].

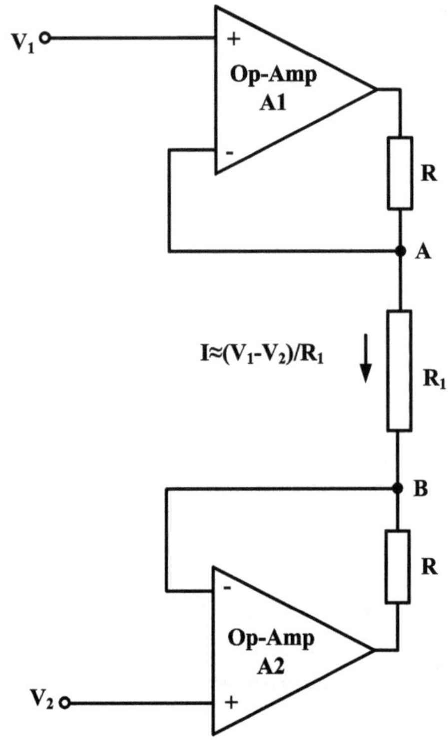

Fig. 2.3 Input stage of conventional three Op-Amp based instrumentation amplifier [13]

2.2.2 *First Generation with Single-Output Structure*

Figure 2.4 shows the schematic diagram of the first generation CMIA based on Op-Amp power supply current sensing technique. It was first introduced by Toumazou and Lidgey [13] and then further expanded and applied by others [14–19]. Op-Amps A_1 and A_2 both operate as voltage buffers. As explained in the Sect. 2.2.1, the current produced across R_1 is equal to $(V_1 - V_2)/R_1$. This current is sensed by current mirrors CM_1 and CM_2. This topology is also called single output topology because the outputs of two current mirrors are combined at node C to make a single output current which is fed to the output stage made of Op-Amp A_3 and resistor R_2. By assuming ideal current mirrors, it provides the output voltage equal to:

$$V_o \approx \frac{R_2}{R_1}(V_1 - V_2) \tag{2.11}$$

Ideally, for common-mode input, no current is produced across R_1 and Op-Amps supply currents do not change, therefore zero voltage is produced at output node.

For differential-mode input, the DC gain of the amplifier is determined by the ratio R_2/R_1. Unlike conventional 3-Op-Amp based IAs where the gain-bandwidth is

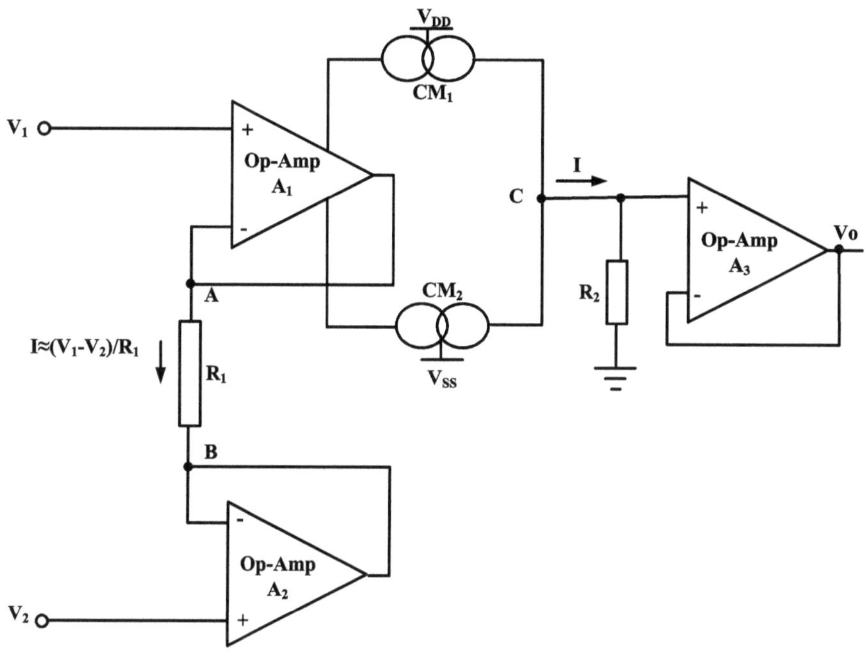

Fig. 2.4 First generation CMIA based on Op-Amp power supply current sensing technique [13]

constant (mentioned in Sect. 1.3, Chap. 1), in the CMIA of Fig. 2.4, resistors are not in the feedback loop, hence the differential gain can be varied by R_2 without affecting circuit bandwidth [13].

In [13], measurement results of the IA of Fig. 2.4 were reported using 4-MHz AD711 Op-Amps together with four-transistor current-mirror constructed from CA3096 BJT arrays. The achieved constant bandwidth differential-mode gain is shown in Fig. 2.5 in which R_1 was kept constant at 100 Ω and R_2 was varied. The plot of CMRR for unity differential-mode gain is shown in Fig. 2.6 which shows a DC value of 80 dB. The quality of Op-Amps and the matching between A_1 and A_2 determines the CMRR performance.

The main problem with the CMIA of Fig. 2.4 is that the finite output impedance of current mirrors affects the output current. To solve this problem, in the topology shown in Fig. 2.7, the output of currents mirrors is kept at virtual ground [14, 15, 17] which cancels out the effect of their finite and unequal output impedance. In this circuit, the output Op-Amp acts as transimpedance amplifier.

In the CMIA shown in Fig. 2.4, if Op-Amps A_1 and A_2 are not identical, a current will follow through R_1 and a non-zero output voltage will be produced for common-mode input [18]. Even if A_1 and A_2 are well matched, current in supply leads is modulated by input common-mode voltage [18]. This condition can be simulated by removing R_1 and applying V_1. Since R_1 is infinite, ideally supply current should not experience any changes. However, supply current shows changes in response to input signal. This unwanted variation can be greatly reduced if the power supply of

Fig. 2.5 Experimental results of differential gain for CMIA of Fig. 2.4 [13]

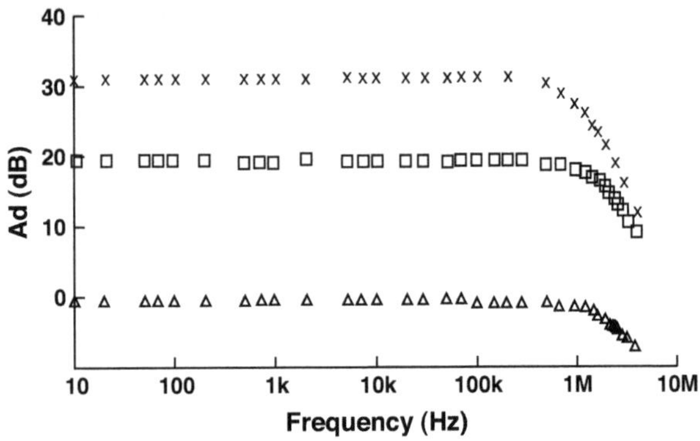

Fig. 2.6 Experimental results of CMRR for unity differential gain for CMIA of Fig. 2.4 [13]

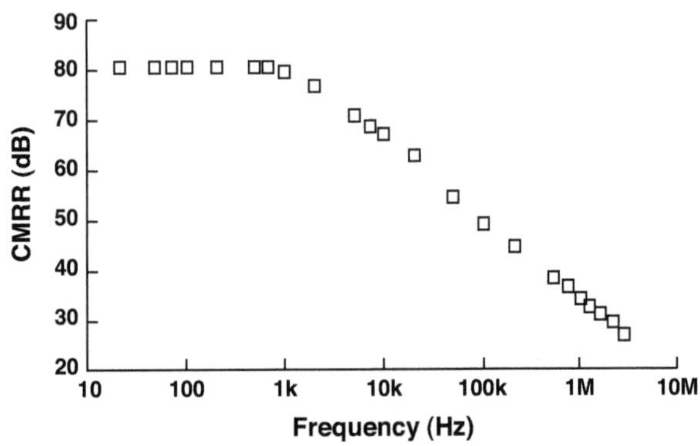

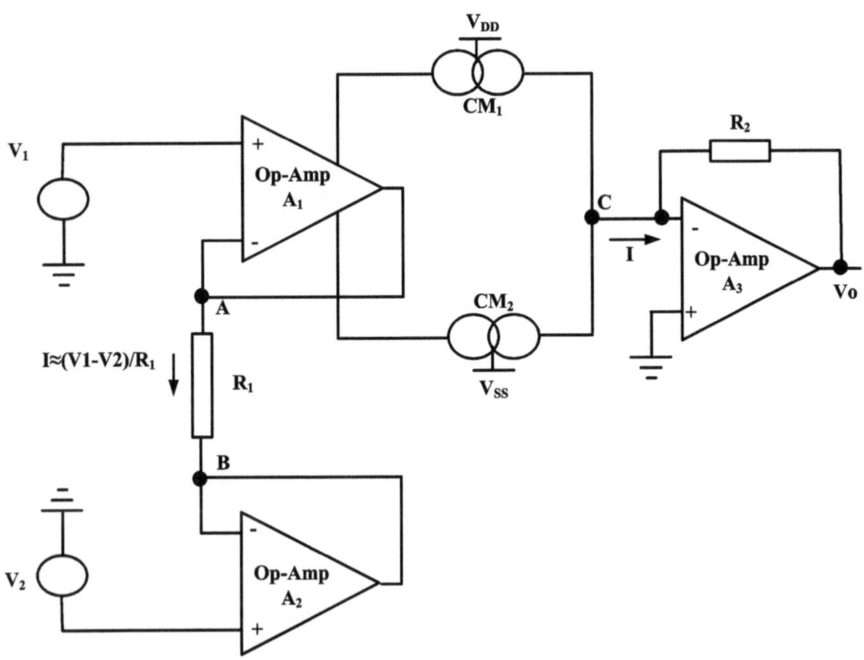

Fig. 2.7 First generation Op-Amp power supply current sensing CMIA with improved current transfer accuracy [17]

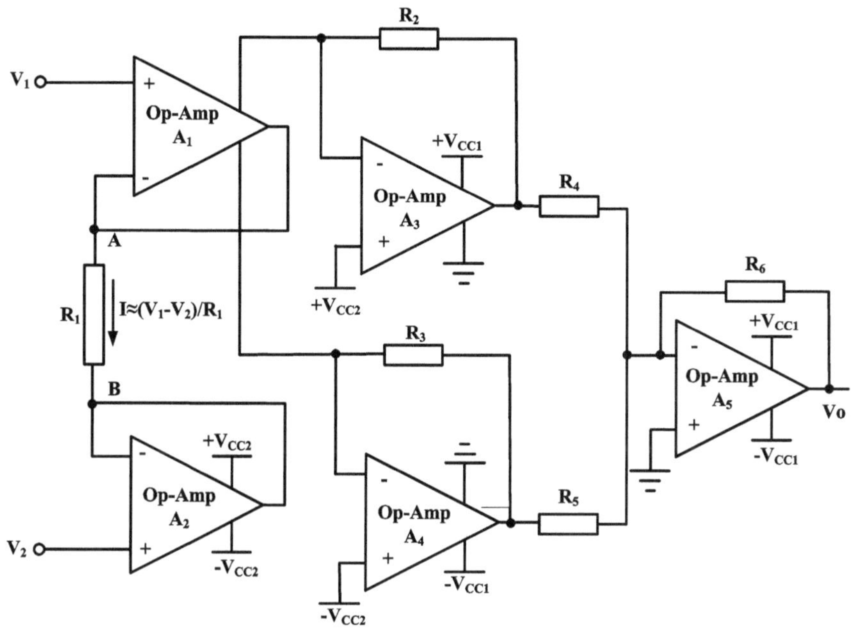

Fig. 2.8 First generation improved Op-Amp power supply current sensing CMIA: $R_1 = R_2 = R_3$; $R_4 = R_5$; $V_{CC1} > V_{CC2}$; $V_O = (R_6/R_1)(V_2 - V_1)$ [18]

A_1 is kept constant [18]. Figure 2.8 shows an improved version in which current mirrors are replaced by a closed loop analog circuitry which keeps the A_1 supply voltages fixed. Power supply changing of A_1 is performed with A_3 and A_4 which are configured as V_{CC} stages. Apparently, A_3 and A_4 require supply voltage larger than V_{CC2}. The summation of output voltages is achieved with A_5 and resistors R_4, R_5 and R_6. Well matching between resistors R_2 and R_3 is required. Resistors R_4 and R_5 must also be well matched. A prototype of CMIA of Fig. 2.8 was constructed using LF351 Op-Amps and CMRR of 60 dB up to 200 kHz was achieved in [18]. In another topology reported in [18], supply voltages are kept constant by means of Zener diodes and supply current is led to output through these diodes. Compared with the five Op-Amp based CMIA shown in Fig. 2.8, in the CMIA based on Zener diodes, the circuit complexity is reduced and resistor matching is not required.

2.2.3 Second Generation with Balanced Structure

In the first generation of Op-Amp supply current sensing CMIA, common-mode signal rejection is mainly performed by input stage which is a differential input voltage controlled current source. The input stage has a single output topology. With the aim of further increasing CMRR, second generation of Op-Amp supply current

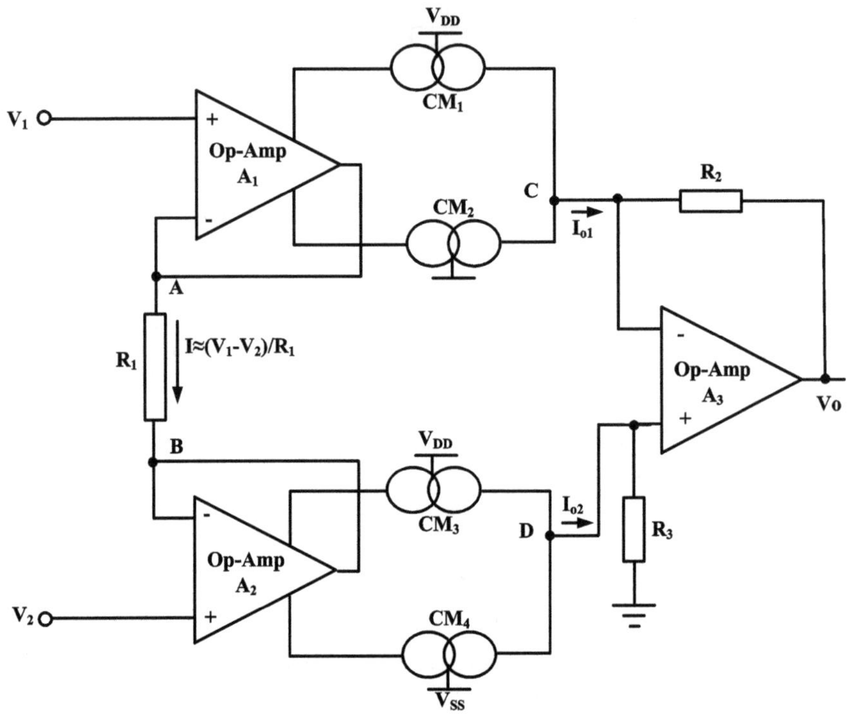

Fig. 2.9 Second generation Op-Amp power supply current sensing CMIA with balanced structure [19]

sensing CMIA was introduced in [19] showing full symmetry. The circuit is reported in Fig. 2.9 in which input stage is changed to dual output. The output differential amplifier provides additional common-mode rejection. However, matching between the two halves of the circuit determines the overall CMRR. In [19], bootstrapping technique is also applied to the second generation CMIA (not shown in Fig. 2.9).

2.2.4 Performance Analysis

2.2.4.1 Performance Analysis of First Generation CMIA

Although the CMIA based on Op-Amp power supply current sensing technique is insensitive to resistor mismatch, however the mismatch between Op-Amps performance parameters has significant effect on its CMRR. In [20], the effect of mismatching between the input Op-Amps main parameters including CMRR, Open Loop Gain (A_o), Gain-Bandwidth product (GBW), and output resistors (r_o) on the CMRR of the first generation Op-Amp power supply current sensing CMIA

(Fig. 2.4) is analyzed. In the analysis reported in [20], mismatch between PSRR and offset voltage is not considered. It is proved that the value of total CMRR can be increased by minimizing the mismatch between input Op-Amps CMRR values and increasing their magnitude. However, in this case the final value of CMRR is determined by open loop gain of input Op-Amps and mismatch between them. The bandwidth of the CMIA voltage gain is also determined by GBW_1 or GBW_2 (the smaller one) cancelling the constant GBW drawback of conventional VMIA [20].

Figures 2.10 and 2.11 show experimental results for differential gain and CMRR reported in [20] which are achieved using CA3096 transistor arrays as current mirrors and different types of Op-Amps for A_1, A_2, and A_3.

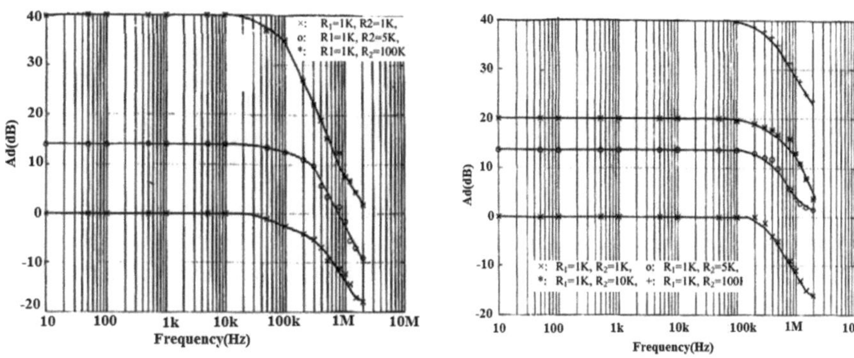

(a) (b)

Fig. 2.10 Differential voltage gain frequency response of first generation CMIA based on supply current sensing technique using (**a**) μA741CN (**b**) TL071CN Op-Amps [20]

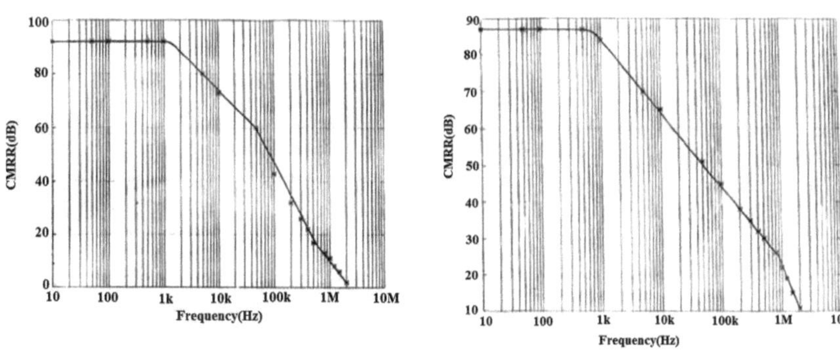

(a) (b)

Fig. 2.11 CMRR frequency response of first generation CMIA based on supply current sensing technique using (**a**) μA741CN (**b**) TL071CN Op-Amps for $R_2/R_1 = 10$ [20]

2.2.4.2 Performance Analysis of Second Generation CMIA

In [21], a detailed analysis of second generation CMIA based on Op-Amp power supply current sensing technique (Fig. 2.9) is given. It is shown that matching between parasitic capacitances and finite output resistances of the Op-Amp internal biasing current sources play a key role in determining CMRR. Other matching conditions are obtained as follows:

$$\mathrm{PSRR}_1 = \mathrm{PSRR}_2 \tag{2.12}$$

$$\lambda_1 - \lambda_2 = \lambda_3 - \lambda_4 \tag{2.13}$$

$$\frac{Ad_1\left(1 \pm \dfrac{1}{2CMRR_1}\right)}{1 + Ad_1\left(1 \mp \dfrac{1}{2CMRR_1}\right)} = \frac{Ad_2\left(1 \pm \dfrac{1}{2CMRR_2}\right)}{1 + Ad_2\left(1 \mp \dfrac{1}{2CMRR_2}\right)} \tag{2.14}$$

where $PSRR_i$, $CMRR_i$ and Ad_i (for i = 1,2) are the power supply rejection ratio, common-mode rejection ratio and differential gain of input Op-Amps, respectively. The term λ_i (for i = 1–4) also denotes the current transfer ratio in the corresponding current mirror. It is proven that achieving an infinite CMRR is not possible. The reason is that as PSRR_1 and PSRR_2 are static variables, so it is practically impossible to realize matching between them.

Figure 2.12 shows CMRR Pspice simulation results of the first generation 5-Op-Amp-based CMIA of Fig. 2.8 and the second generation CMIA of Fig. 2.9 both with applying bootstrapping technique and without applying it using Op-Amp 741 transistor level model. As it is clear, CMRR performance shows significant improvement in second generation topology especially with bootstrapping technique.

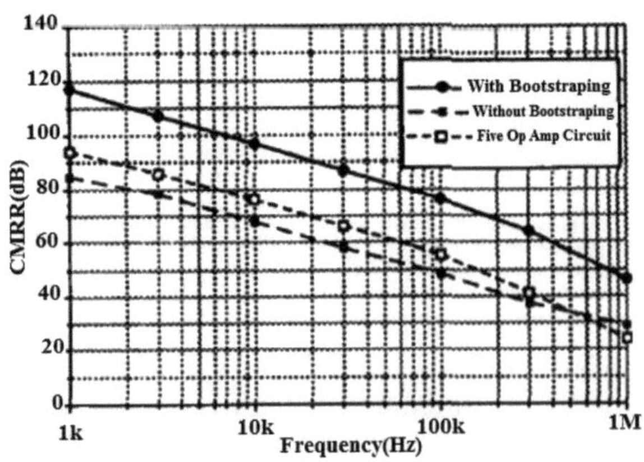

Fig. 2.12 CMRR frequency performances of first generation CMIA of Fig. 2.8 and second generation CMIA of Fig. 2.9 [19]

2.2.5 Comparison Between Different Generations of CMIA Based on Op-Amp Power Supply Sensing Technique

The main difference between various types of Op-Amp supply current sensing CMIA is on their output stage. In CMIA shown in Fig. 2.4, all the three Op-Amps operate as voltage buffers. Therefore, by the assumption that current mirrors −3 dB bandwidth is much higher than Op-Amps GBW, its frequency performance is very high and close to the used Op-Amps unity gain bandwidths. However, to increase current transfer accuracy, current mirrors with a very high output impedance is required.

In the CMIA of Fig. 2.7, the two input Op-Amps operate as voltage buffers and the output Op-Amp is in negative feedback loop performing as transimpedance amplifier. Therefore, in this case bandwidth will be determined by output stage. However, the output of current mirrors are kept at ground potential and high output impedance current mirrors are not required.

In the second generation CMIA shown in Fig. 2.9, similar to the CMIA of Fig. 2.7, the output stage determines the overall bandwidth. Similar to the CMIA of Fig. 2.4, it also requires high output impedance current mirrors. The superiority of second generation CMIA over first generation one is that the output of its first stage is in a fully differential configuration and as a result of differential action the effect of input Op-Amps mismatches is greatly reduced. However, due to the used output stage, second generation CMIA is sensitive to mismatches between the used resistors.

2.3 Conclusion

In this chapter, we have discussed Op-Amp supply current sensing technique and various implementations of CMIA based on this technique. It is also shown that compared to conventional three Op-Amp Voltage-mode IA (Sect. 1.3, Chap. 1), CMIA based on Op-Amp supply current sensing technique shows some advantages. First, to achieve high CMRR only good Op-Amp matching is required without the need for ideal Op-Amp and strict resistor matching. Second, the gain can be easily controlled by only one resistor. Third, its bandwidth is high and independent of gain. However, requiring high supply voltage is its main demerit.

It is mentioned that in the second generation CMIA, CMRR mainly is determined by the matching conditions between input Op-Amps and current-mirrors, and not by their performance.

References

1. Haslett J. W., Rao M K N (1979) A high quality controlled current source. IEEE Transactions Instrumentation Measurement, 28(2):132–140.
2. Nordholt E. H.(1982) Extending Opamp capabilities by using power-supply. IEEE Transactions on Circuits and Systems, 29(6):411–414.
3. Toumazou C., Lidgey F. J. (1985) Floating-impedance convertors using current conveyors. Electronic Letters, 21(15):64–642.
4. Toumazou C., Lidgey F. G. (1986) Universal active filter using current conveyors. Electronics Letters, 22(12): 662–664.
5. Toumazou C., Lidgey F. G. (1987) Wide-band precision rectification. IEE Proceedings 134(1), DOI: https://doi.org/10.1049/ip-g-1.1987.0002.
6. Wilson B., Lidgey F. G., Toumazou C. (1988) Current activated analogue signal processing circuits. Proc. IEEE International symposium on circuits and systems, June 1988.
7. Wilson B. (1985), Floating FDNR employing a new CCII-conveyor implementation. Electronic Letters, 21:996–997.
8. Sharif-Bakhtiar M., Aronhime P. (1978) A current conveyor realization using operational amplifiers, International Journal of Electronics, 45: 283–288.
9. Annema A. J., Nauta B., Langevelde R.V., Tuinhout H. (2005) Analog circuits in ultra-deep-submicron CMOS. IEEE Journal of Solid State Circuits, 40:132–143.
10. Fayomi C. J. B., Sawan M., Roberts G.W. (2004) Reliable circuit techniques for low-voltage analog design in deep submicron standard CMOS: a tutorial. Analog Integrated Circuits and Signal Processing, 39:21–38.
11. Su W. J., Lidgey F. J., Porta S., Zhu Q. S. (1994) Analysis of IC op-amp power-supply current sensing. IEEE International Symposium on Circuits and Systems, 1994.
12. Lidgey F. J., Su W. J.(1994) Improvements to op-amp power-supply current sensing technique. Electronic Letters, 30(19):1567–1568.
13. Toumazou C., Lidgey F. J. (1989) Novel current-mode instrumentation amplifier. Electronic Letters, 25(3):228–230.
14. Douglas E. L. et al. (2004) A Low-voltage current-mode instrumentation amplifier designed in a 0.18-micron CMOS technology. Proceedings of Canadian Conference on Electrical and Computer Engineering, 2004.
15. Harb A., Sawan M. (1999) New low-power low-voltage high-CMRR CMOS instrumentation amplifier.Proceedings of IEEE International Symposium on Circuits and Systems, 1999.
16. Prior C. A., Vieira F. C. B., Rodrigues C. R. (2006), Instrumentation amplifier using robust rail-to-rail operational amplifiers with gm control. Proceedings of IEEE Int. Midwest Symposium on Circuits and Systems, 2006.
17. Li J. T., Pun S. H., Mak P. U., Vai M. I. (2008) Analysis of op-amp power-supply current sensing current-mode instrumentation amplifier for biosignal acquisition system. 30th Annual International IEEE EMBS Conference, 2008.
18. Zhu Q. S., Lidgey F. J., Hunt M. A. V.(1992) Improved wide-band, high CMRR instrumentation amplifier. Clin. Phys. Physiol. Phys. Meas., 13(Suppl. A):51–55.
19. Zhu Q. S., Lidgey E J., Su W. J.(1993) High CMRR, second generation current-mode instrumentation amplifiers. Proceedings of IEEE Int. Symposium on Circuits and Systems, 1993.
20. Azhari S. J., Fazlalipoor H. (2009) CMRR in voltage-op-amp-based current-mode instrumentation amplifiers (CMIA). IEEE Transaction on Instrumentation and Measurement, 58(3):563–569.
21. Su W. J., Lidgey F. J. (1995) Common-mode rejection ratio in current-mode instrumentation amplifiers.Analog Integrated Circuits Signal Processing, 7(3):257–260.

Chapter 3
Current-Mode Wheatstone Bridge

3.1 Traditional Voltage-Mode Wheatstone Bridge

3.1.1 *Introduction to Traditional Voltage-Mode Wheatstone Bridge*

3.1.1.1 Voltage-Mode Wheatstone Bridge with Voltage Excitation

Voltage-Mode Wheatstone Bridge (VMWB) is traditionally used for accurately measuring small resistance variations. It is interesting to note that the circuit was invented by S.H. Christie in 1833 [1]. However, since Charles Wheatstone was the first one who used the circuit to measure resistances, it is known today as Wheatstone Bridge. Recently, it has found many applications in temperature, pressure and resistive measurements [2–16]. Figure 3.1 shows VMWB circuit [2, 3] that represents the fully differential version of voltage divider and consists of four resistors R_1–R_4 and a constant voltage source V_{ref}. A voltage detector measures the voltage difference between nodes A and B which is expressed as:

$$V_{out} = V_{ref}\left(\frac{R_3}{R_1 + R_3} - \frac{R_4}{R_2 + R_4}\right) \tag{3.1}$$

The balancing condition $R_3/R_1 = R_4/R_2$ gives $V_{out} = 0$. Variation of one or more resistors from its initial value causes output voltage to change which is interpreted as a magnitude variation in the measured variable. As it is seen from Eq. (3.1), the output voltage of the Bridge is directly proportional to V_{ref}. Therefore, the measurement accuracy is limited by that of V_{ref}.

Table 3.1 shows different configurations of Wheatstone Bridge and related output voltage. Based on the number of sensors used, Wheatstone Bridges are usually arranged in a quarter configuration which includes only one sensor, a half

G. Ferri et al., *Current-Mode Instrumentation Amplifiers*, Analog Circuits and Signal Processing, https://doi.org/10.1007/978-3-030-01343-1_3

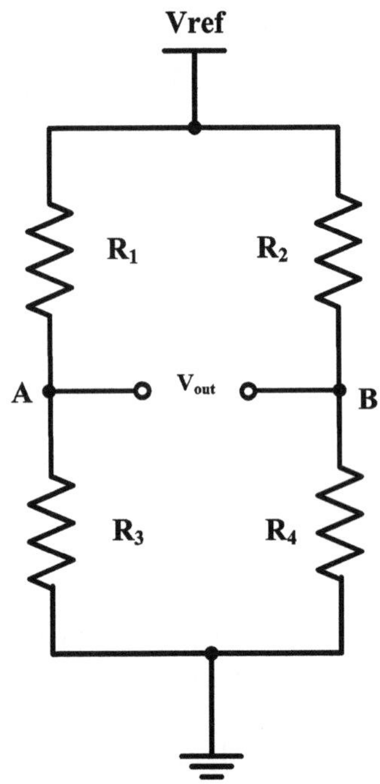

Fig. 3.1 Traditional voltage-mode Wheatstone bridge circuit [3]

configuration where two equal sensors are used or a full configuration having four sensors, two-by-two equal [2, 3]. For example, in the full configuration we have:

$$R_1 = R_4 = R_0 \mp \Delta R \tag{3.2}$$

$$R_2 = R_3 = R_0 \pm \Delta R \tag{3.3}$$

Inserting Eq. (3.2) and Eq. (3.3) into Eq. (3.1), the output voltage is found as:

$$V_{out} = V_A - V_B = \pm \frac{\Delta R}{R_0} V_{ref} \tag{3.4}$$

As it is seen from Eq. (3.4), increasing V_{ref} results in a proportional increase in the output voltage. However, it also results in higher power consumption and sensor self-heating errors.

A possible definition of the sensitivity of a Bridge is the ratio of maximum output voltage change to the excitation voltage V_{ref}. For example if V_{ref} is 10 V and the maximum output voltage change is 10 mV, the Bridge sensitivity is found as 1 mV/V [3].

Table 3.1 Different configurations of traditional VMWB with voltage excitation [3]

Configuration	Schematic	Output voltage
(a) **A quarter configuration with a single sensor**	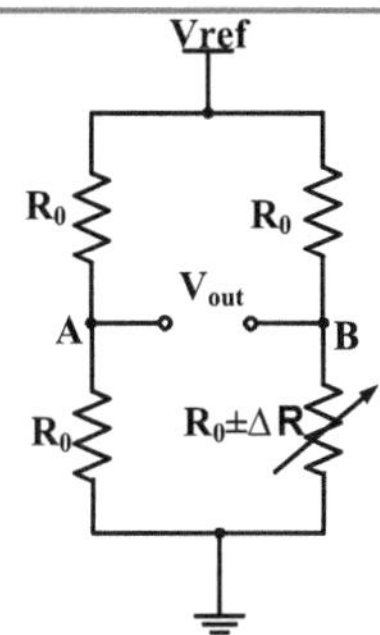	$V_{out} = \frac{V_{ref}}{4}\frac{\pm\Delta R}{R_0 \pm \Delta R/2}$
(b) **Half configuration with two equal sensors**	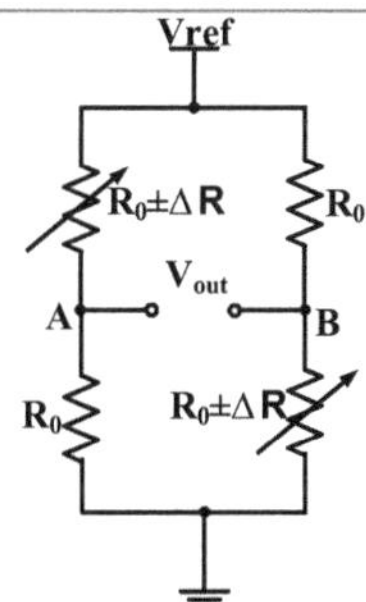	$V_{out} = \frac{V_{ref}}{2}\frac{\pm\Delta R}{R_0 \pm \Delta R/2}$
(c) **Half configuration with two different sensors**	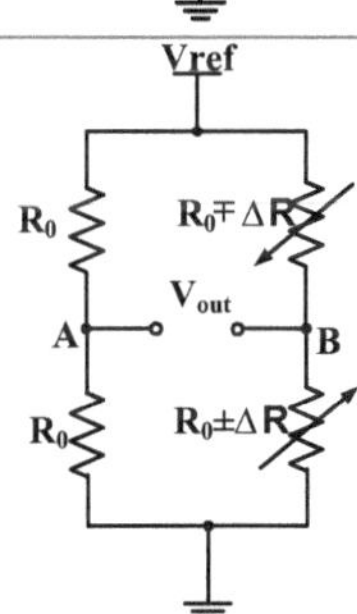	$V_{out} = \frac{V_{ref}}{2}\frac{\pm\Delta R}{R_0}$
(d) **Full configuration with four sensors, two-by-two equal**	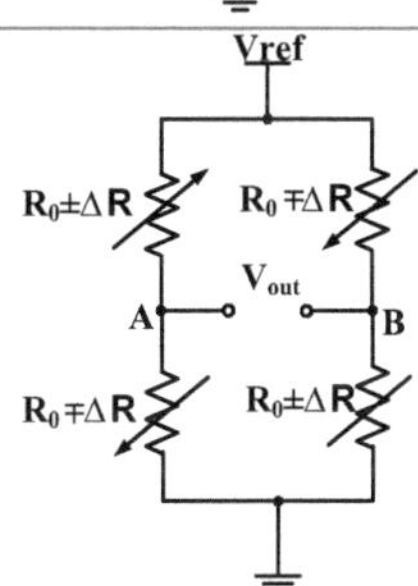	$V_{out} = V_{ref}\frac{\pm\Delta R}{R_0}$

3.1.1.2 Voltage-Mode Wheatstone Bridge with Current Excitation

In the case where the Bridge is placed in a distance from excitation source, it is usually biased by a constant current source [3]. By this approach the effect of wiring resistance on the measurement accuracy is reduced effectively; therefore, expensive cabling is not required. Table 3.2 shows different configurations of Wheatstone Bridge excited by a constant current source I_{ref}. As it is seen, except for single sensor configuration case, all the other configurations are inherently linear. It should also be noted that, compared to voltage excited Bridge, current excitation requires a current source implementation (not detailed below) which increases circuit complexity.

3.1.2 Read-Out Circuits for Voltage-Mode Wheatstone Bridge

3.1.2.1 Time-Based Approach

Figure 3.2 shows the concept of time-based approach for VMWB read-out circuit described in [4] which includes one differential sensor resistance, a capacitor, a hysteresis comparator, a time to digital convertor, three switches and a voltage source V_{ref}. In this approach, the discharge times of capacitor C over two sensing resistors $R_1 = R_0(1 + \Delta R)$ and $R_2 = R_0(1 - \Delta R)$ which are named t_1 and t_2 respectively, are utilized. The discharge times t_1 and t_2 are measured in four steps by employing S_0, S_1 and S_2 switches as follows.

In the first step two switches S_1 and S_2 are opened while S_0 is closed, so capacitor C is charged to V_{ref}. In the second step, switches S_0 and S_2 are opened and S_1 closed. Thus capacitor C discharges over R_1 until its voltage reaches to the value V_{TH} that triggers the comparator. The time interval between closing S_1 and triggering comparator, t_1, is measured by the time-to-digital convertor. After having triggered the comparator, the third step starts where S_1 and S_2 are opened, S_0 is closed and capacitor C is charged to V_{ref} again. In the final step, capacitor C is discharged over R_2 (by closing S_2 and opening S_1 and S_0) until the comparator is triggered again. The time interval between closing of S_2 and triggering comparator, t_2, is measured. The relation between the value of ΔR and the measured time intervals (t_1 and t_2) is stated in [4] as:

$$\frac{t_1 - t_2}{t_1 + t_2} = \frac{R_1 - R_2}{R_1 + R_2} = \frac{R_0(1+\Delta R) - R_0(1-\Delta R)}{R_0(1+\Delta R) + R_0(1-\Delta R)} = \frac{\Delta R}{R_o} \tag{3.5}$$

The advantage of time-based approach is the reduction of the number of resistors in the Bridge. However, on/off resistances of switches, delay time of comparator, errors related to time to digital convertor such as quantization error and gain error limit the accuracy of this approach.

Table 3.2 Different configurations of traditional VMWB with current excitation [3]

Configuration	Schematic	Output voltage
(a) A quarter configuration with a single sensor	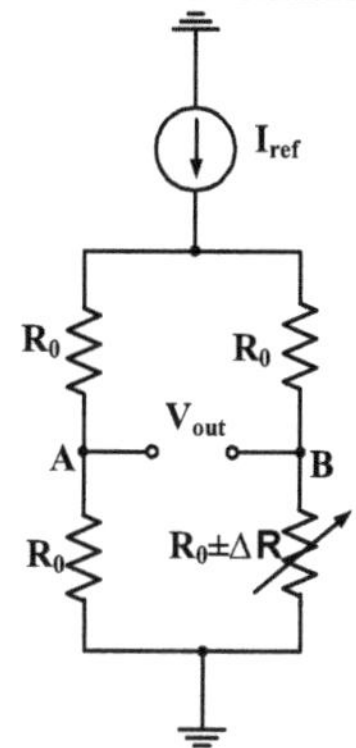	$V_{out} = \frac{I_{ref}R_0}{4}\frac{\pm\Delta R}{R_0 \pm \Delta R/4}$
(b) Half configuration with two equal sensors	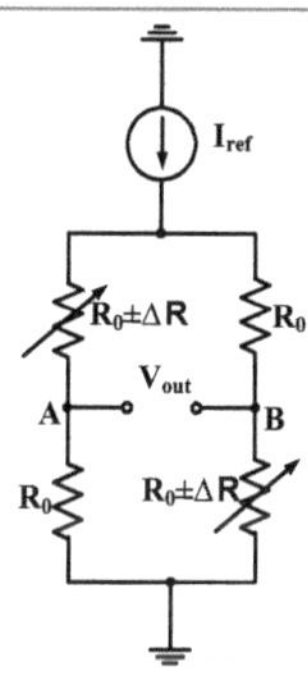	$V_{out} = \frac{I_{ref}}{2}(\pm\Delta R)$
(c) Half configuration with two different sensors	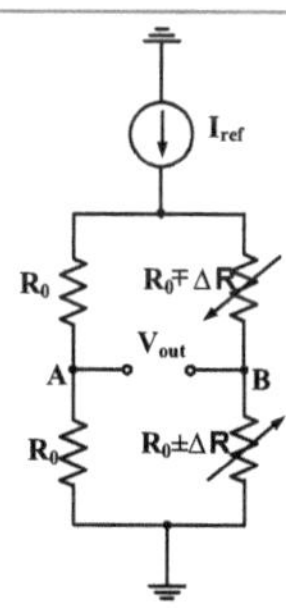	$V_{out} = \frac{I_{ref}}{2}(\pm\Delta R)$
(d) Full configuration with four sensors, two-by-two equal	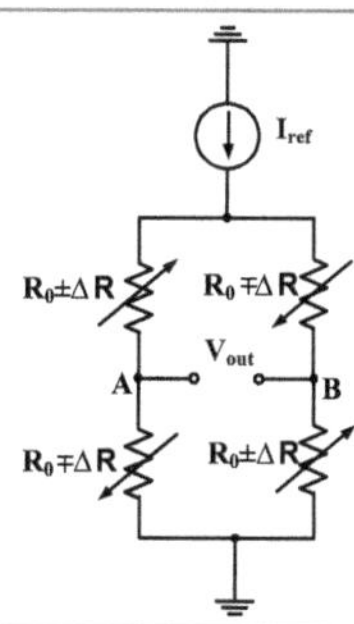	$V_{out} = I_{ref}(\pm\Delta R)$

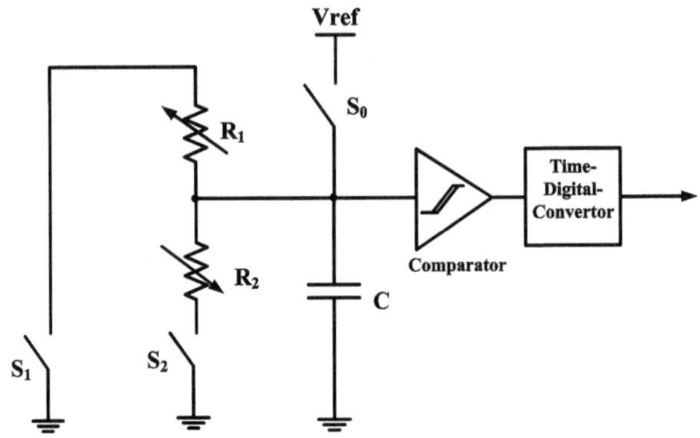

Fig. 3.2 Time-based approach for VMWB read-out circuit [4]

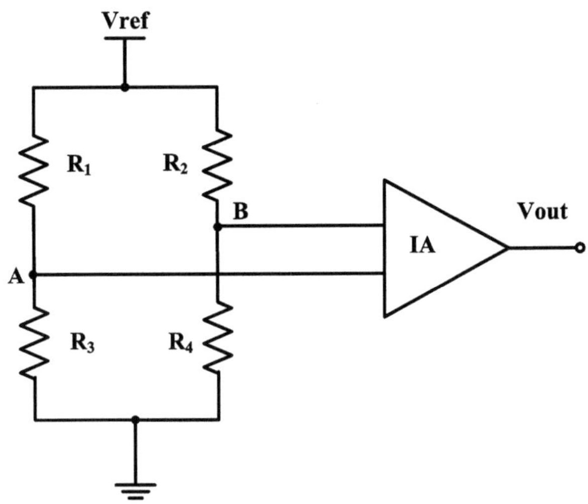

Fig. 3.3 Voltage-based approach for VMWB read-out circuit [4]

3.1.2.2 Voltage-Based Approach

In the voltage-based approach, the output voltage of Wheatstone Bridge is typically amplified by an IA as shown in Fig. 3.3. Adjustable high differential gain, large input impedance, low offset voltage and large CMRR are main characteristics of an IA. The relation between input and output voltages of IA is given as:

$$V_{out} = A\left(V_A - V_B\right) \tag{3.6}$$

where V_A and V_B are the input voltages of IA and A is its gain.

INA118 is an example of a high-quality monolithic IA from Burr-Brown/Texas Instruments with a low offset voltage of 50 μV and a high CMRR of 110 dB [5]. Its gain is adjustable by means of an external resistor.

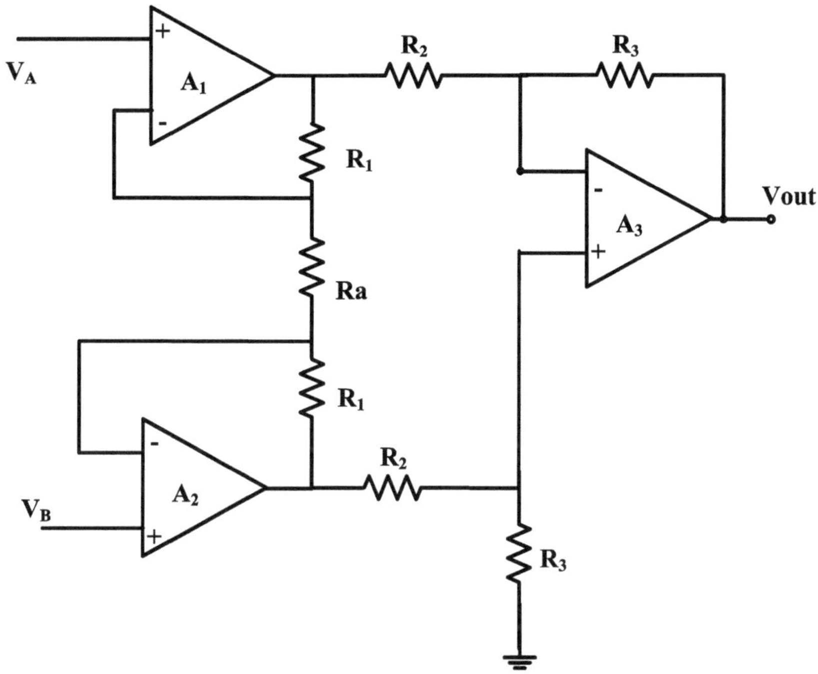

Fig. 3.4 Conventional three Op-Amp based IA

Figure 3.4 shows schematic of conventional IA made of three Op-Amps and seven resistors (refer to Sect. 1.3, Chap. 1). In this circuit, the voltage across R_a is forced to be equal to input voltage and the overall gain is:

$$A = \left(1 + \frac{2R}{R_a}\right)\frac{R_3}{R_2} \tag{3.7}$$

In this circuit, the matching between resistors determines the final value of CMRR. As a general estimation, 1% mismatch between resistors limits the CMRR to 40 dB while for 0.1%, the CMRR is limited to 60 dB [5].

3.1.3 Linearization of Voltage-Mode Wheatstone Bridge

As it is clear from Tables 3.1 and 3.2, the relation between output voltage and ΔR for configurations with single sensor is not linear. This non-linearity can be compensated using software or circuit based techniques. In [3], the negative feedback approach is used for linearization. Figure 3.5 shows a circuit example employing this method. The negative feedback, employed through an Op-Amp, provides a

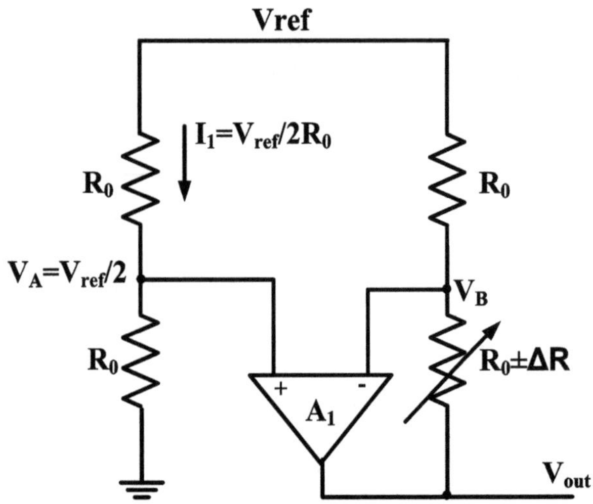

Fig. 3.5 VMWB linearization using negative feedback technique [3]

constant current equal to $V_{ref}/2R_0$ in the varying element by keeping $V_B = V_A$. A straightforward analysis performed in [3] shows that the output voltage is calculated as Eq. (3.8) which is a linear function of ΔR:

$$V_{out} \approx -\frac{\Delta R}{2R_0} V_{ref} \tag{3.8}$$

As other examples, in [17–19] the nonlinearity is compensated by modulating the Bridge excitation source in a feedback loop. In [20], linearity is achieved using a specifically designed trans-impedance amplifier. In [21] negative feedback technique and in [22, 23] positive feedback technique are used for Bridge linearization.

3.2 Current-Mode Wheatstone Bridge

3.2.1 Introduction to Current-Mode Wheatstone Bridge

Figure 3.6 shows the principle of an alternative kind of Bridge, named the Current-Mode Wheatstone Bridge (CMWB) introduced in [24]. The CMWB is the current-mode counter part of conventional VMWB and has been developed using the circuit duality concept. As Fig. 3.6 shows, the circuit includes two resistors excited by a current source I_{ref}. The resistors are connected at one end and forced to have equal

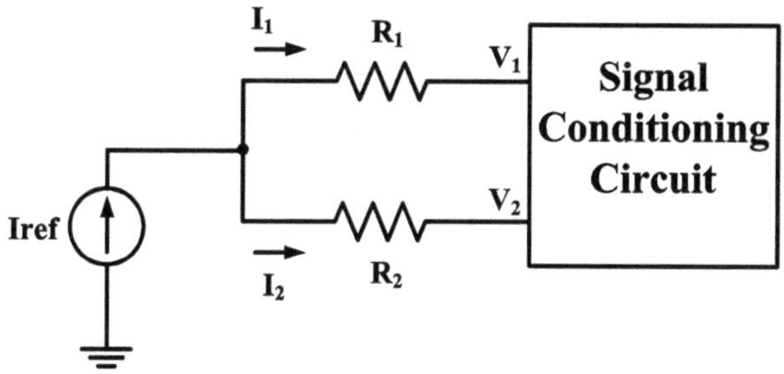

Fig. 3.6 Current-mode Wheatstone bridge principle [24]

and constant voltage (V_c) at the other end. Therefore, the following conditions must be fulfilled by the signal conditioning circuit:

$$V_1 = V_2 = V_c \tag{3.9}$$

$$I_{ref} = I_1 + I_2 \tag{3.10}$$

In the circuit of Fig. 3.6, I_1 and I_2 can be expressed as:

$$I_1 = \frac{R_2}{R_1 + R_2} I_{ref} \tag{3.11}$$

$$I_2 = \frac{R_1}{R_1 + R_2} I_{ref} \tag{3.12}$$

Assuming $R_1 = R_0 \pm \Delta R$ and $R_2 = R_0 \mp \Delta R$, gives the difference between I_1 and I_2 as a linear functions of ΔR:

$$I_1 - I_2 = \Delta I = \pm \frac{\Delta R}{R_0} I_{ref} \tag{3.13}$$

The main problem of the CMWB of Fig. 3.6 is related to its large common-mode currents which are calculated as:

$$I_{CM} = \frac{I_1 + I_2}{2} = \frac{I_{ref}}{2} \tag{3.14}$$

To reduce errors due to such large common-mode input currents, it is necessary to implement high CMRR signal conditioning circuits. Another possible solution is to cancel the effect of common-mode input currents by adding two current sinks or resistive dividers as shown in Fig. 3.7. The used current sinks and resistor dividers drain $I_{ref}/2$ and prevent entering the common-mode current to signal conditioning

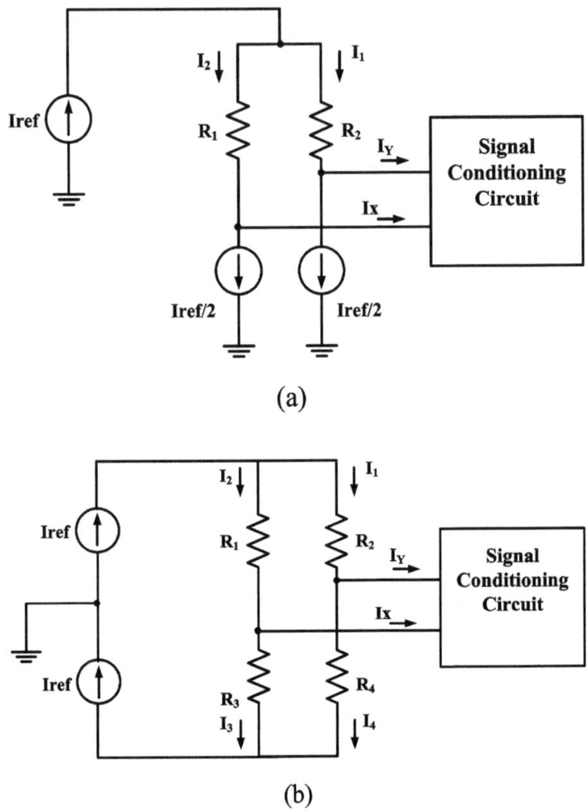

Fig. 3.7 Common-mode current cancellation in CMWB using (**a**) current sink (**b**) resistive dividers [24]

circuit. Unfortunately, these solutions destroy the circuit simplicity which was the main advantage of CMWB over conventional voltage-mode one. Compared to VMWB which uses a voltage source to excite the Bridge, a CMWB requires an accurate current source with complicated implementation.

The main advantage of a CMWB is that the resistor elements are equipotential at both ends, so superposition of multiple sensors is very straightforward as shown in Fig. 3.8. In this case by assuming $I_{refi} = I_{ref}$ and $R_{1i} = R_0 \pm \Delta R$ and $R_{2i} = R_0 \mp \Delta R$, for $i = 1,\ldots n$, we have:

$$I_1 - I_2 = \Delta I = \pm n \frac{\Delta R}{R_0} I_{ref} \tag{3.15}$$

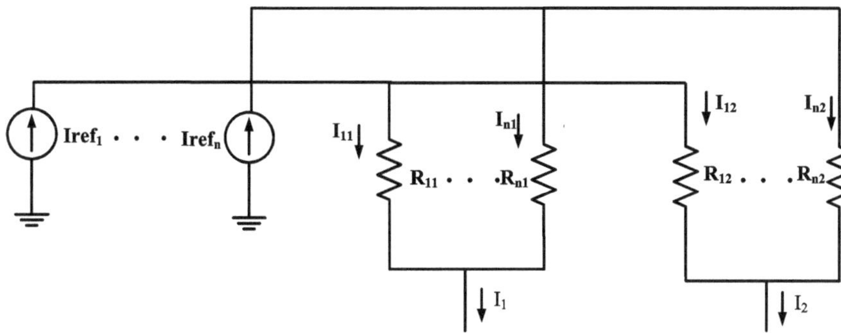

Fig. 3.8 Superposition of multiple sensors in CMWB [24]

3.2.2 Linearization of Current-Mode Wheatstone Bridge

In the cases where only one sensor is used in CMWB, the output of Fig. 3.6 is a nonlinear function of ΔR:

$$\Delta I = I_1 - I_2 = \pm \frac{\Delta R}{2R_0 \pm \Delta R} I_{ref} \tag{3.16}$$

Figure 3.9 shows a CCII-based method reported in [24] to compensate this nonlinearity. In this figure, I_x is given by:

$$I_x = \frac{V_x}{R_2} - I_{ref} \tag{3.17}$$

For CCII, $I_y = 0$, therefore $I_1 = I_{ref}$ and V_x is obtained as:

$$V_x = V_y = I_1 R_1 = I_{ref} R_1 \tag{3.18}$$

By inserting Eq. (3.18) into Eq. (3.17), we get:

$$I_x = \frac{\mathrm{R1}}{R_2} I_{ref} - I_{ref} \tag{3.19}$$

Inserting $R_1 = R_0 \pm \Delta R$, $R_2 = R_0$, into Eq. (3.19), I_{out} is achieved as:

$$I_{out} = \pm \frac{\Delta R}{R_0} I_{ref} \tag{3.20}$$

Eq. (3.20) is achieved by assuming an ideal CCII. Unfortunately, the non-idealities of the used CCII deteriorate the linearity of I_{out}.

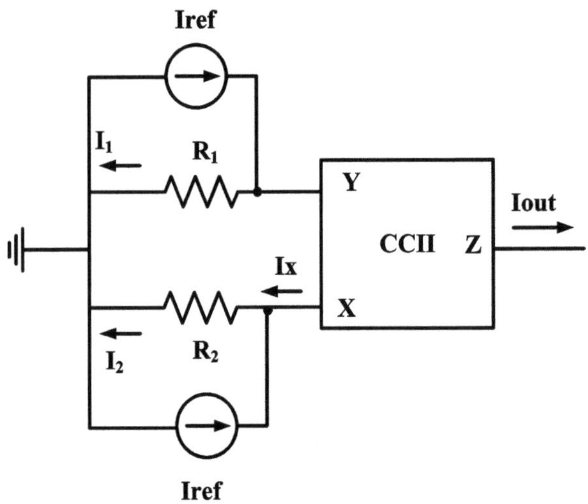

Fig. 3.9 Nonlinearity compensation in CMWB [24]

3.2.3 Read-Out Circuits for Current-Mode Wheatstone Bridge

3.2.3.1 OFCC-Based Read-Out Circuit

Figure 3.10 shows the read-out circuit reported in [24] which uses Operational Floating Current Conveyor (OFCC) as active building block and six resistors. The OFCC has two input and three output ports. X is a low-impedance current input port, Y is a high-impedance voltage input port, W is low-impedance voltage output port and Z+ and Z− are high-impedance current output ports. There is a voltage tracking between Y and X nodes and a current tracking between W and Z nodes. In an ideal OFCC, $I_Y = 0$, $V_X = V_Y$, $V_W = Z_t I_x$, $I_{Z+} = -I_{Z-} = I_W$ where $Z_t = \infty$. The OFCC is designed to be used in closed loop configuration with current feedback from node W to node X [25, 26]. In Fig. 3.10, OFCC$_1$ and resistors R_3, R_4 form a negative feedback loop producing output currents of:

$$I_{z1+} = -I_{z1-} = \left(1 + \frac{R_4}{R_3}\right) I_{R1} \tag{3.21}$$

Similarly, the output currents of OFCC$_2$ are:

$$I_{z2+} = -I_{z2-} = \left(1 + \frac{R_4}{R_3}\right) I_{R2} \tag{3.22}$$

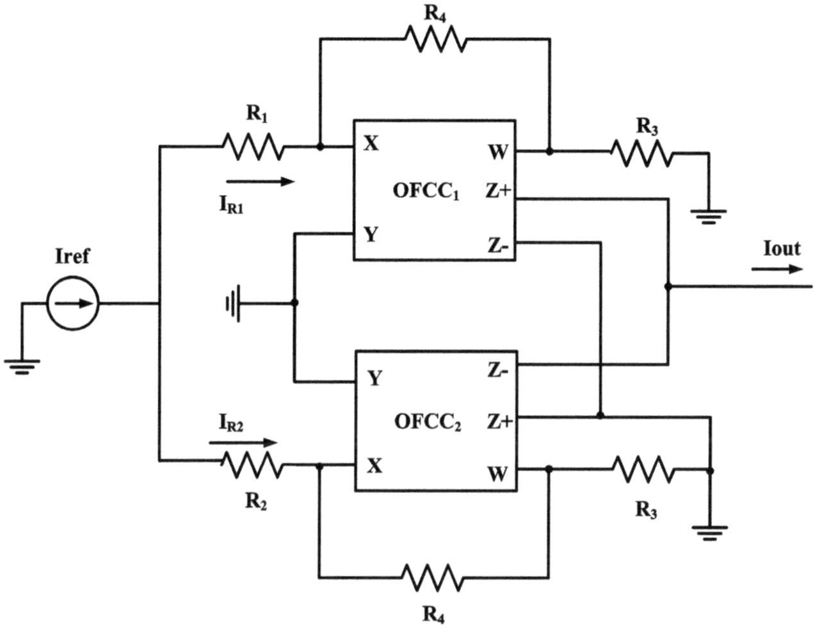

Fig. 3.10 OFCC-based read-out circuit [24]

As it is seen from Fig. 3.10, the Z^+ output of $OFCC_1$ is tied to Z^- output of $OFCC_2$ making a current subtraction node. Therefore I_{out} is calculated as:

$$I_{out} = I_{z1+} - I_{z2-} = \left(1 + \frac{R_4}{R_3}\right)\left(I_{R1} - I_{R2}\right) \tag{3.23}$$

By noting that $V_{X1} = V_{X2} = 0$, I_{R1} and I_{R2} can be expressed as:

$$I_{R1} = \frac{R_2}{R_1 + R_2} I_{ref} \tag{3.24}$$

$$I_{R2} = \frac{R_1}{R_1 + R_2} I_{ref} \tag{3.25}$$

By inserting I_{R1} and I_{R2} into Eq. (3.23), and assuming $R_1 = R_0 \pm \Delta R$ and $R_2 = R_0 \mp \Delta R$, I_{out} is achieved as:

$$I_{out} = \left(1 + \frac{R_4}{R_3}\right)\left(\frac{R_1 - R_2}{R_1 + R_2}\right) = \left(1 + \frac{R_4}{R_3}\right)\left(\frac{\Delta R}{R_0}\right) \tag{3.26}$$

The main feature of this approach is its high accuracy because due to negative feedback loop, output current is independent of intrinsic resistance at X node. However, in the case of a single sensor, the circuit requires linearization. In [25], a possible linearization of OFCC based read-out circuit is also discussed.

3.2.3.2 CCII Based Read-Out Circuit

Figure 3.11 shows a very simple approach reported in [24] to develop a read-out circuit. The circuit is based on negative and positive second generation current conveyors ($CCII^-$ and $CCII^+$). The Z terminals of CCIIs are tied together to make a current subtraction node. The operation of this circuit is very simple. As the voltage at X terminals of CCIIs are at ground, I_{R1} and I_{R2} are expressed as:

$$I_{R1} = \frac{R_2}{R_1 + R_2} I_{ref} \tag{3.27}$$

$$I_{R2} = \frac{R_1}{R_1 + R_2} I_{ref} \tag{3.28}$$

By assuming ideal CCIIs, the output current is calculated as:

$$I_{out} = I_{z1} - I_{z2} = I_{R1} - I_{R2} = \frac{R_1 - R_2}{R_1 + R_2} I_{ref} \tag{3.29}$$

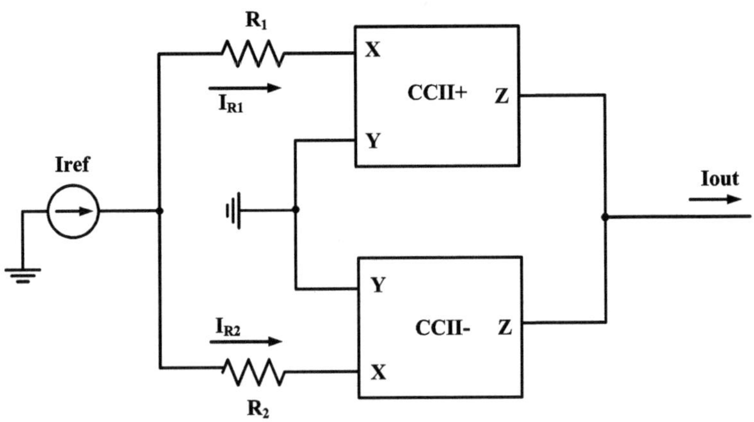

Fig. 3.11 CCII-based read-out circuit [24]

According to Eq. (3.29), n case of two sensors $R_1 = R_0 \pm \Delta R$ and $R_2 = R_0 \mp \Delta R$, I_{out} is a linear function of ΔR:

$$I_{out} = \pm \frac{\Delta R}{R_0} I_{ref} \tag{3.30}$$

while for single varying element, the output is a nonlinear function of ΔR. In this circuit, the accuracy is limited by the X terminals intrinsic resistances.

3.2.3.3 COA-Based Read-Out Circuit

The idea of using Current Operational Amplifier (COA) to implement read out circuits of CMWB is based on the general concept discussed in [27–29] stating that each voltage processing circuit can be turned into a current processing circuit by replacing voltage Op-Amps with their equivalent current-mode elements, i.e. COAs (Fig. 3.12) and interchanging the input and output terminals of the circuit. Figure 3.12a shows the symbolic representation of COA as the current-mode counter part of voltage Op-Amp. It exhibits ideally zero input resistance ($r_{in} = 0$), infinite open loop current gain ($A_0 = \infty$) and infinite output resistance ($r_o = \infty$). It has one input and two outputs. The relation between input and output currents can be written as:

$$i_o^+ = i_o^- = A_0 i_{in} \tag{3.31}$$

Figure 3.13a shows the conventional Op-Amp based amplifier which is suitable for VMWB. Figure 3.13b shows a current-mode read-out circuit obtained from Fig. 3.13a by applying the duality concept, based on COAs.

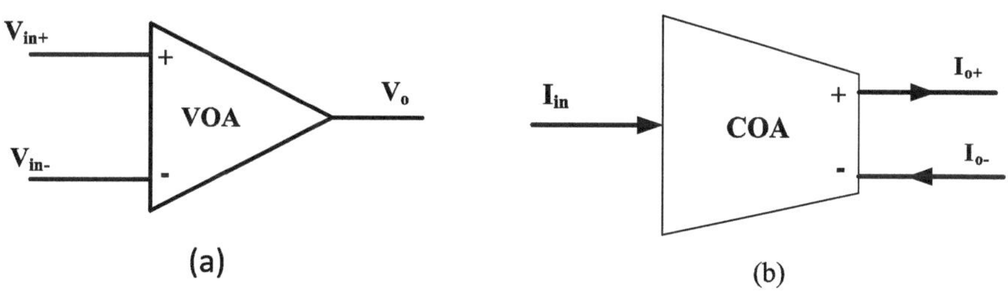

Fig. 3.12 Symbolic representation of (**a**) conventional voltage Op-Amp (**b**) COA as its current-mode adjoint [30]

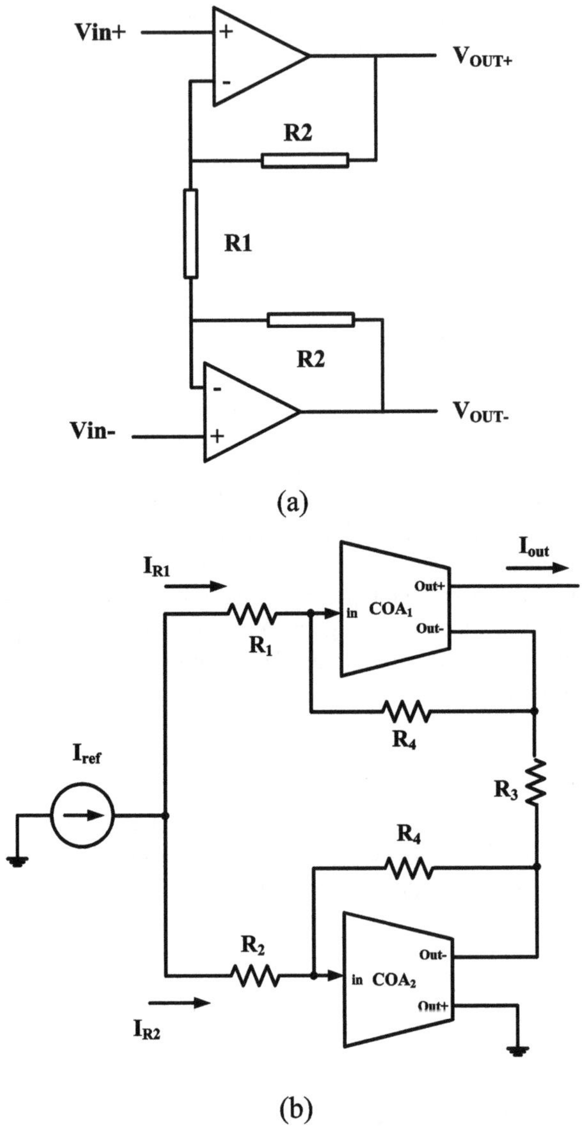

Fig. 3.13 Read-out circuit for (**a**) conventional VMWB (**b**) CMWB [30]

In Fig. 3.13b, by assuming ideal COAs with infinite gain and zero input impedance, the input voltages of COA_1 and COA_2 are zero. Therefore I_{R1} and I_{R2} are calculated approximatively as:

$$I_{R1} = \frac{R_2}{R_1 + R_2} I_{ref} \tag{3.32}$$

$$I_{R2} = \frac{R_1}{R_1 + R_2} I_{ref} \tag{3.33}$$

The output current can be expressed as:

$$I_{out} = \frac{R_4}{R_3}(I_{R1} - I_{R2}) + I_{R1} = \frac{R_4}{R_3}\left(\frac{R_2 - R_1}{R_1 + R_2}\right)I_{ref} + \frac{R_2}{R_1 + R_2}I_{ref} \tag{3.34}$$

By replacing $R_1 = R_0 \pm \Delta R$ and $R_2 = R_0 \mp \Delta R$ into Eq. (3.34) and arranging it, I_{out} is achieved as:

$$I_{out} = I_{ref}\left(\frac{1}{2} + \frac{R_4}{R_3}\right)\frac{\pm\Delta R}{R_0} + \frac{1}{2}I_{ref} \tag{3.35}$$

Although the output current is a linear function of ΔR, but a large value of common-mode output equal to $I_{ref}/2$ appears at output node which must be cancelled. A new approach is introduced in [30] where the common-mode component of this circuit is cancelled by replacing dual output COA with multiple output one. However, in case of single sensor applications, this configuration also requires linearization.

3.2.3.4 Current Mirror-Based Read-Out Circuit

Figure 3.14 shows the CMWB read-out circuit which is based on three current mirrors realized by M_1-M_6 transistors and an Op-Amp [31]. Here R_1 and R_2 are sensors excited by the constant current source I_{ref}. One of their terminals is connected together while the others are forced, by Op-Amp, to have equal voltages, i.e. $V_X = V_Y$. The currents in R_1 and R_2 are respectively found as:

$$I_1 = \frac{R_2}{R_1 + R_2}I_{ref} \tag{3.36}$$

$$I_2 = \frac{R_1}{R_1 + R_2}I_{ref} \tag{3.37}$$

From Fig. 3.14, by assuming current mirrors M_1-M_3 and M_5-M_6 to have unity gain, currents I_1 and I_2 are subtracted at output node to produce I_{out} as:

$$I_{out} = I_1 - I_2 \tag{3.38}$$

By assuming $R_1 = R_0 \pm \Delta R$ and $R_2 = R_0 \mp \Delta R$ and inserting I_1 and I_2 from Eq. (3.36) and Eq. (3.37) into Eq. (3.38), I_{out} is given by:

$$I_{out} = \frac{\pm\Delta R}{R_0}I_{ref} \tag{3.39}$$

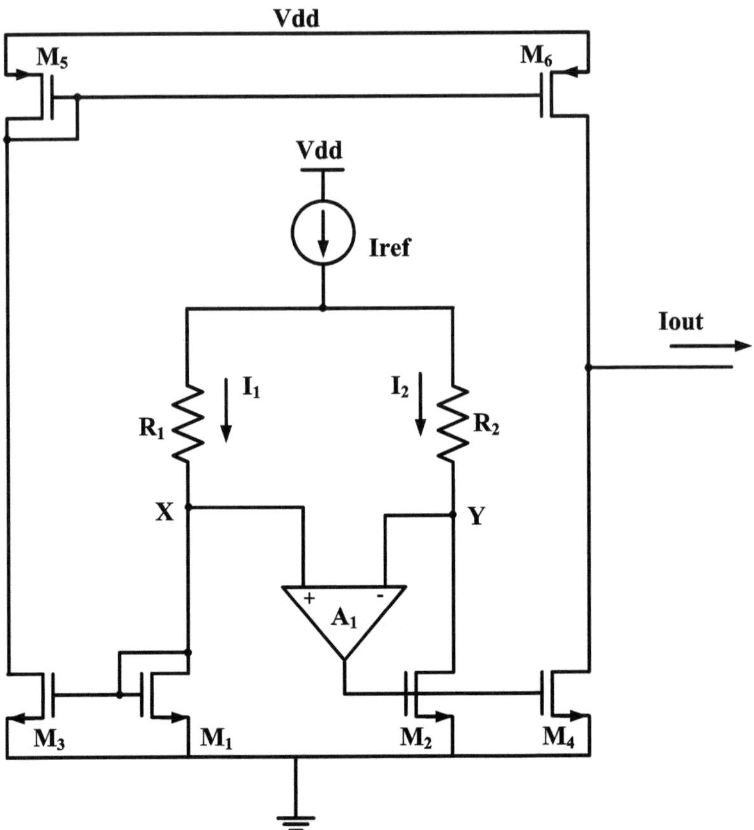

Fig. 3.14 Current mirror based read-out circuit for CMWB [31]

If only one varying element is used, i.e. $R_2 = R_0 \pm \Delta R$ and $R_1 = R_0$, the output current becomes a nonlinear function of ΔR, as follows:

$$I_{out} = \frac{\pm \Delta R}{2R_0 + \Delta R} I_{ref} \tag{3.40}$$

In this case linearization can be performed by a small modification, as is shown in Fig. 3.15 [31]. In this circuit the current in varying element, R_2, is kept constant through I_{ref} resulting $V_Y = (R_0 \pm \Delta R)I_{ref}$. By considering the fact that $V_Y = V_X$, the current in R_1 will be:

$$I_1 = \frac{R_0 \pm \Delta R}{R_0} I_{ref} \tag{3.41}$$

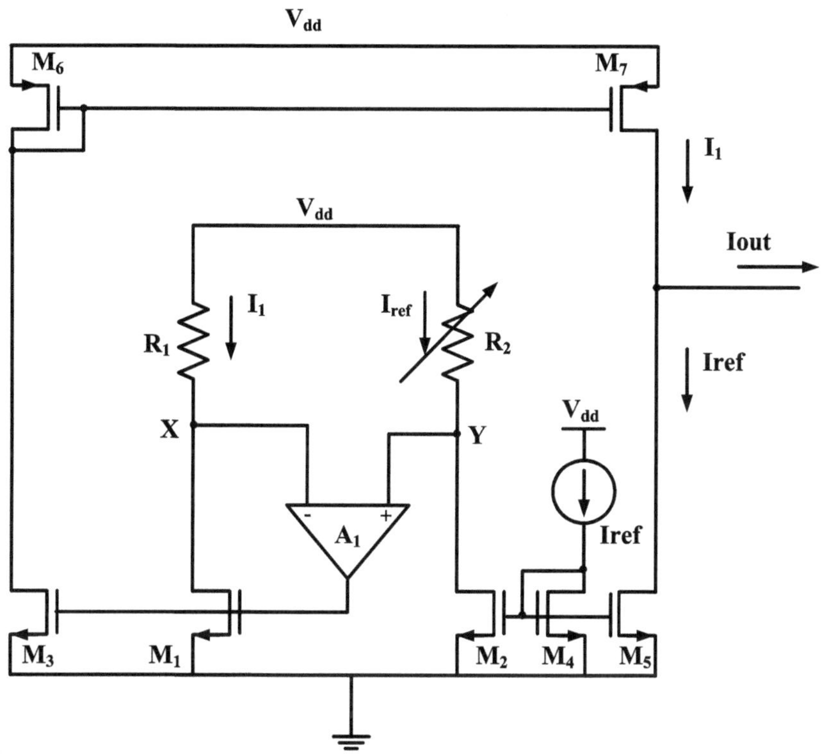

Fig. 3.15 Linearization technique for Fig. 3.13 [31]

Inserting $I_2 = I_{ref}$ and I_1 from Eq. (3.41) into Eq. (3.38), the output current becomes a linear function of ΔR, as follows:

$$I_{out} = I_1 - I_{ref} = \frac{\pm \Delta R}{R_0} I_{ref} \tag{3.42}$$

3.2.3.5 CDTA Based Read-Out Circuit

Figure 3.16 shows a read-out circuit based on dual output Current Differencing Transconductance Amplifier (CDTA) reported in [32]. A dual output CDTA is defined by the following matrix [32]:

$$\begin{bmatrix} V_p \\ V_n \\ I_z \\ I_{x1} \\ I_{x2} \end{bmatrix} = \begin{bmatrix} 00000 \\ 00000 \\ 1(-1)000 \\ 0000g_{m1} \\ 0000g_{m2} \end{bmatrix} \begin{bmatrix} I_p \\ I_n \\ V_{x1} \\ V_{x2} \\ V_z \end{bmatrix} \tag{3.43}$$

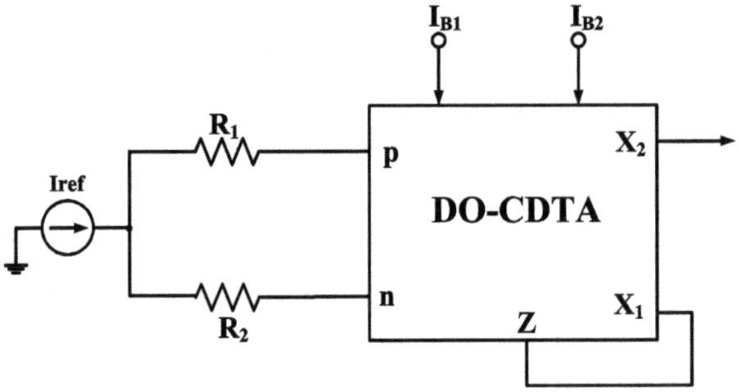

Fig. 3.16 Dual-output CDTA based read-out circuit for C MWB [32]

where for BJT realization of CDTA:

$$g_{m1} = \frac{I_{B1}}{2V_T}, \quad g_{m2} = \frac{I_{B2}}{2V_T} \tag{3.44}$$

and I_{B1} and I_{B2} are CDTA bias currents and V_T is thermal voltage. In Fig. 3.16, $V_n = V_p = 0$, so I_n and I_p are calculated as:

$$I_n = \frac{R_2}{R_1 + R_2} I_{ref} \tag{3.45}$$

$$I_p = \frac{R_1}{R_1 + R_2} I_{ref} \tag{3.46}$$

Using Eq. (3.45) and Eq. (3.46), I_z is found as:

$$I_z = I_n - I_p = \frac{R_2 - R_1}{R_1 + R_2} I_{ref} \tag{3.47}$$

The current at X_1 terminal is $I_{x1} = I_z = -g_{m1} V_z$, therefore V_z is written as:

$$V_z = \frac{I_{x1}}{g_{m1}} = \left(\frac{R_1 - R_2}{R_1 + R_2}\right) \frac{I_{ref}}{g_{m1}} \tag{3.48}$$

By assuming $R_1 = R_0 \pm \Delta R$ and $R_2 = R_0 \mp \Delta R$ and using $I_{out} = I_{x2} = -g_{m2} V_Z$, I_{out} is found as:

$$I_{out} = -\frac{g_{m2}}{g_{m1}} \left(\frac{R_1 - R_2}{R_1 + R_2}\right) I_{ref} = -\frac{I_{B2}}{I_{B1}} \left(\frac{\Delta R}{R_0}\right) I_{ref} \tag{3.49}$$

The advantage of the circuit of Fig. 3.16 is that its gain is electronically controllable by I_{B2} and I_{B1}. However, in case of a single sensor, it requires linearization. In [33] a similar read-out circuit based on single output CDTA is introduced. It employs an extra current controllable CCII (CCCII).

3.2.3.6 CDBA and CCCII-Based Read-Out Circuit

Figure 3.17 shows the read-out circuit introduced in [33]. It uses one Current Differencing Buffered Amplifier (CDBA) and two Current Controlled CCII (CCCII). For CDBA we have $V_p = V_n = 0$, $I_z = I_p - I_n$ and $V_w = V_z$. The operation of CCCII is similar to CCII except that its resistances at X terminal (r_x) is electronically controllable. In the circuit of Fig. 3.16, for $CCCII_1$ and $CCCII_2$ the r_x are controlled by I_{B1} and I_{B2} as:

$$r_{x1} = \frac{I_{B1}}{2V_T} \tag{3.50}$$

$$r_{x2} = \frac{I_{B2}}{2V_T} \tag{3.51}$$

In $CCCII_1$, V_{ref} applied to Y terminal is transferred to X terminal and converted to a current proportional to r_{x1}. This current which is equal to $I_{ref} = V_{ref}/r_{x1}$ is used to excite R_1 and R_2. As the voltage at p and n terminals of CDBA is equal to zero, for I_{R1} and I_{R2} we have, respectively:

$$I_{R1} = I_p = \frac{R_2}{R_1 + R_2} I_{ref} \tag{3.52}$$

$$I_{R2} = I_n = \frac{R_1}{R_1 + R_2} I_{ref} \tag{3.53}$$

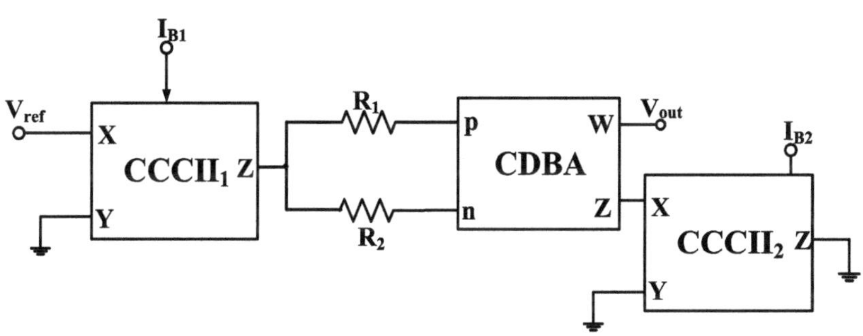

Fig. 3.17 CDBA based read-out circuit for CMWB [33]

The current at Z terminal of CDBA is equal to $I_Z = I_p - I_n$ and a voltage equal to $V_z = I_z r_{x2}$ is produced at Z terminal of CDBA. As a result of voltage tracking between Z and W terminals of CDBA, V_{out} is found as:

$$V_{out} = I_Z r_{x2} = \left(\frac{R_1 - R_2}{R_1 + R_2}\right)\frac{r_{x2}}{r_{x1}} V_{ref} \tag{3.54}$$

For $R_1 = R_0 \pm \Delta R$ and $R_2 = R_0 \mp \Delta R$, from Eq. (3.54), V_{out} will be:

$$V_{out} = \left(\pm\frac{\Delta R}{R_0}\right)\frac{r_{x2}}{r_{x1}} V_{ref} \tag{3.55}$$

In case of one varying element, the transfer function of Eq. (3.55) is nonlinear. As a clear advantage, the realization of circuit of Fig. 3.17 is simple because it does not require any reference current source implementation.

3.2.3.7 Op-Amp Based Read-Out Circuit

In [24] a simple read-out circuit based on Op-Amps and resistors is reported. In this circuit which is shown in Fig. 3.18, R_1 and R_2 are sensors excited by I_{ref}. Op-Amps A_1-A_2 and resistors R_3 make trans-impedance amplifiers which convert I_{R1} and I_{R2} to voltage signals. A differential amplifier made of A_3 and resistors R_4-R_5 are used to amplify the difference of output voltage of A_1 and A_2. Again by assuming $R_1 = R_0 \pm \Delta R$ and $R_2 = R_0 \mp \Delta R$, the output voltage is found as:

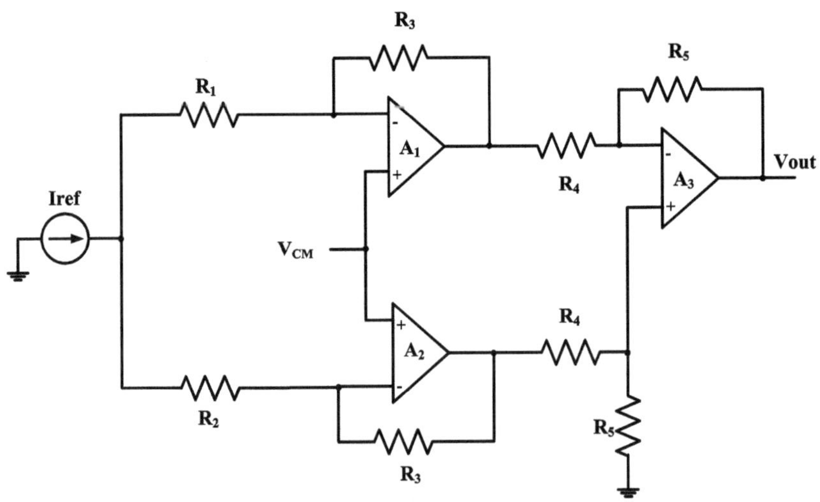

Fig. 3.18 Op-Amp based read-out circuit for CMWB [24]

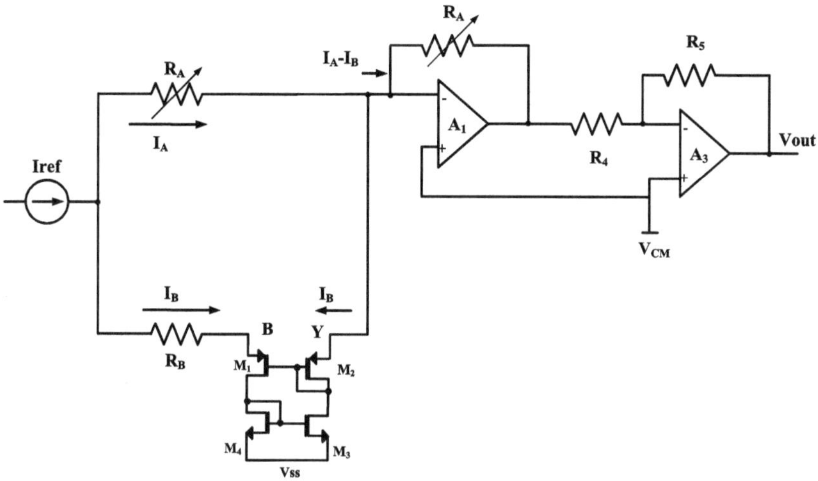

Fig. 3.19 Op-Amp based read-out circuit for CMWB reported in [34]

$$V_{out} = \frac{R_3 R_5}{R_4}\left(\frac{R_1 - R_2}{R_1 + R_2}\right) I_{ref} = \pm \frac{R_3 R_5}{R_4} \frac{\Delta R}{R_0} I_{ref} \tag{3.56}$$

Output voltage is nonlinear if only one sensor is used. In addition, the operation of circuit is highly dependent on matching between resistors. The high number of resistors used in this circuit is its other major drawback. There are also high values of DC offset voltage at Op-Amps inputs since it is equal to $R_0 I_{REF}/2$. Figure 3.18 shows a solution to remove this offset voltage in which two current sources equal to $I_{REF}/2$ are connected to Op-Amps inputs. These current sources must be exactly equal.

Another Op-Amp based read-out circuit is reported in [34] and shown in Fig. 3.19 where a current negative impedance converter (INIC) is used to cancel the effect of offset current $I_{ref}/2$. In ideal case where there is perfect matching between transistor pairs, the voltage at Y and B nodes will be equal allowing R_A and R_B to perform as parallel resistors. INIC inverts the current of R_B (named I_B) and transfers it to Y node where it is subtracted from current of R_A (named I_A). This current subtraction cancels the effect of offset current $I_{ref}/2$ at the Op-Amps inputs. The last stage consisting of A_3 and resistors R_4-R_5 is used to amplify the output of A_1.

3.2.3.8 VCII-Based Read-Out Circuit

In [35] second generation voltage conveyor (VCII) is used to design CMWB read-out circuit. VCII is conceived by applying the duality concept on CCII. Figure 3.20 shows the symbolic representation of VCII and its function is characterized by Eq. (3.57):

$$i_x \approx \pm \beta i_y, V_z = \alpha V_x, V_Y = 0 \tag{3.57}$$

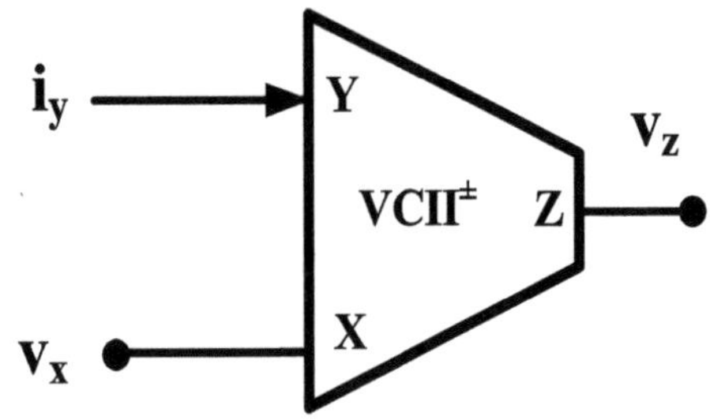

Fig. 3.20 Symbol representation of VCII [35]

where β and α are current gain between Y and X terminals and voltage gain between X and Z terminals, respectively. Current and voltage gains are close to unity and can be expressed as:

$$\beta = 1 \pm \varepsilon_i ; \alpha = 1 \pm \varepsilon_v \tag{3.58}$$

where $\varepsilon_i \ll 1$ and $\varepsilon_v \ll 1$ are error terms. There are two types of VCII identified by the direction of i_x. In VCII$^-$ we have $i_x = -\beta i_y$ and in VCII$^+$ we have $i_x = +\beta i_y$. In VCII Y and Z terminals exhibit low impedance (ideally zero) and X terminal has high impedance (ideally ∞).

Figure 3.21 shows the VCII based read-out circuit in which R_1 and R_2 are sensors and R_3 is used to control gain. Due to low impedance at Y terminals (which is ideally zero), the voltage at Y terminals are assumed to be zero. Therefore, the currents at R_1 and R_2 are:

$$I_1 = \frac{R_0 \mp \Delta R}{R_1 + R_2} I_{ref} ; I_2 = \frac{R_0 \pm \Delta R}{R_1 + R_2} I_{ref} \tag{3.59}$$

Currents I_2 is injected into Y terminals of VCII$_2^+$. The X terminal of VCII$_2^+$ is connected to the Y terminal of VCII$_1^+$. This connection allows a current subtraction between I_1 and I_{x2} resulting I_{Y1} as:

$$I_{Y1} = I_1 - \beta_2 I_2 \tag{3.60}$$

I_{Y1} is transferred to X terminal of VCII$_1^+$ and converted to voltage by R_3:

$$V_{X1} = \beta_1 \left(I_1 - \beta_2 I_2 \right) R_3 \tag{3.61}$$

Using Eq. (3.58) and Eq. (3.61), the output voltage produced at VCII$_1^+$ Z terminal is:

$$V_{out} = \beta_1 \left(I_1 - \beta_2 I_2 \right) \alpha_1 R_3 \tag{3.62}$$

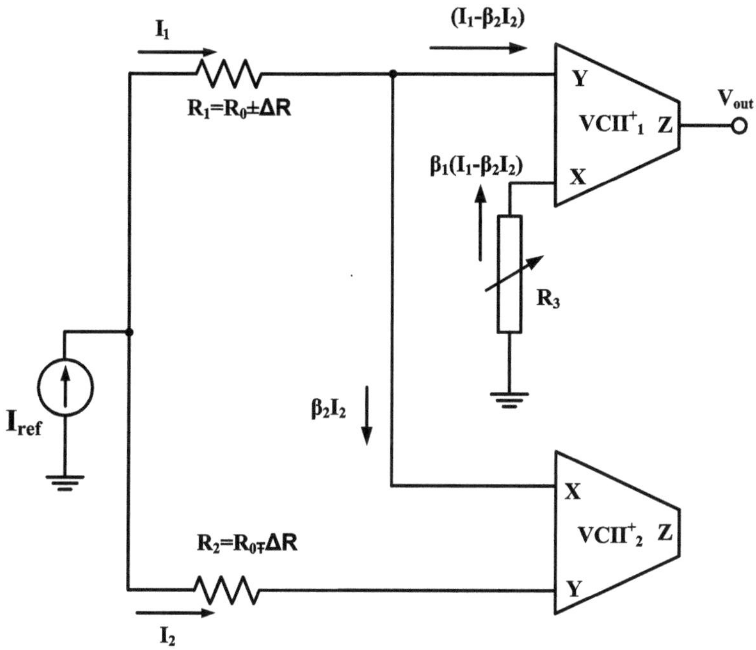

Fig. 3.21 VCII based CMWB read-out circuit for two sensor applications [35]

Inserting Eq. (3.58) and Eq. (3.59) into Eq. (3.62) V_{out} results to be a linear function of ΔR:

$$V_{out} \cong \frac{\pm\Delta R}{R_0}\alpha_1 R_3 I_{ref} \tag{3.63}$$

In case of single sensor applications, the output is nonlinear function of ΔR. Another circuit reported in [35] and shown in Fig. 3.22 is suitable for these applications. For $R_2 = R_0$ and $R_1 = R_0 \pm \Delta R$, its output voltage is:

$$\mathrm{V_{out}} = (\beta_1\alpha_1 - 1)\alpha_2\beta_2 \mathrm{R_3 I_{ref}} + \frac{\pm\Delta R}{\mathrm{R_0}}\alpha_2\beta_2\mathrm{R_3 I_{ref}} \tag{3.64}$$

According to Eq. (3.64) an offset voltage is also produced which is negligible because the condition $\beta_1\alpha_1 \cong 1$ is always met.

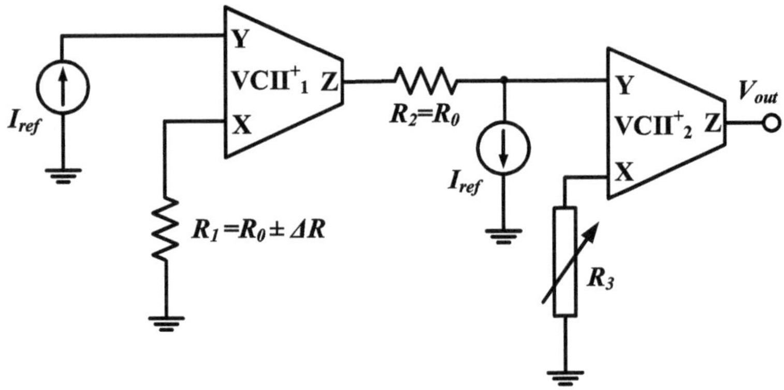

Fig. 3.22 VCII based CMWB read-out circuit for one sensor applications [35]

3.3 Mixed-Mode Wheatstone Bridge and Read-Out Circuits

Mixed-mode Wheatstone Bridge have been employed to take the advantages of both current-mode and voltage-mode signal processing. Figure 3.23 shows two different realizations of mixed-mode Wheatstone Bridge and read-out circuits which employ Op-Amps and $CCII^+$ [36]. Similar to CMWB, the outputs of Mixed-Mode Wheatstone Bridge are current signals, so enabling it to use the advantages of current-mode signal processing. One terminal of sensors is tied to V_{ref} and the other ends are kept equipotential by means of the read-out circuit. Therefore, the resistors operate as parallel resistors. In these circuits R_3 is the varying element. In Fig. 3.23a, for $CCII^+$, we have $V_x = V_y = 0$ and $I_z = I_x$. Therefore, $I_1 = V_{ref}/R_1$ and $I_3 = V_{ref}/R_3$ and V_{out} is expressed, for high Op-Amp gains, as [36]:

$$V_{out} = R_2\left(\frac{1}{R_3} - \frac{1}{R_1}\right)V_{ref} \tag{3.65}$$

By inserting $R_1 = R_0$ and $R_3 = R_0 + \Delta R$ into Eq. (3.65), and assuming $\Delta R \ll R_0$, V_{out} will be:

$$V_{out} = -\frac{R_2}{R_0}V_{ref}\frac{\Delta R}{R_0} \tag{3.66}$$

A similar equation holds for Fig. 3.23b with opposite sign. The output voltage in Mixed-Mode approach is 4 times larger than the voltage based approach and two times larger than the current based approach [36]. Another advantage of Mixed-Mode approach is that multiple sensors can be easily inserted. The number of resistors is also reduced, when compared to traditional voltage-mode approach. Current source (through an I_{ref}) implementation is not also required resulting in a simple realization.

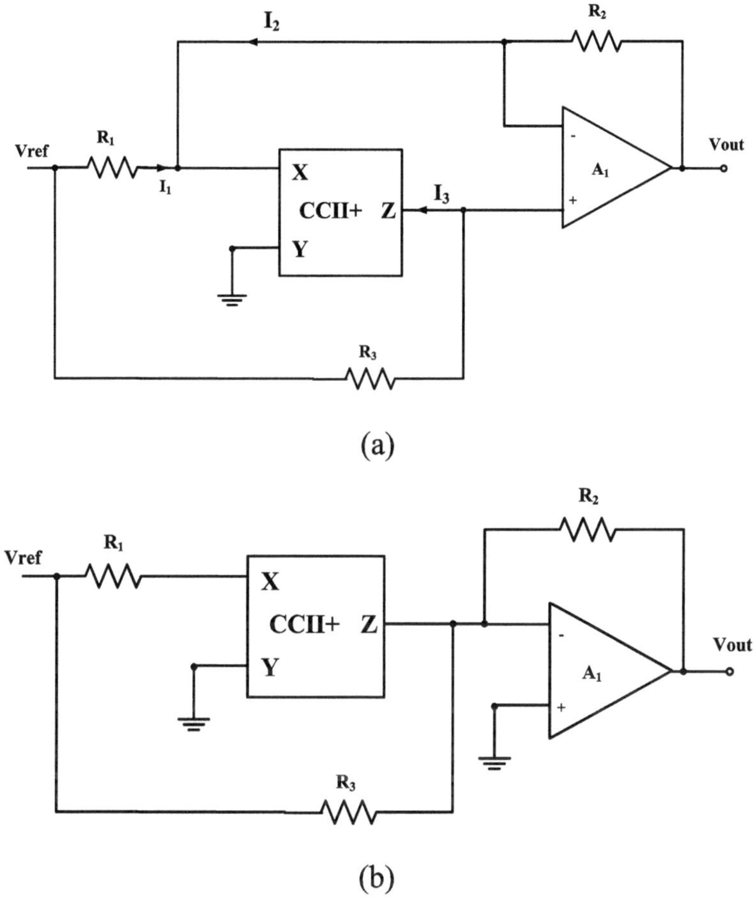

Fig. 3.23 Mixed-mode based read-out circuit (**a**) inverting output (**b**) non-inverting output [36]

References

1. Ekelof S. (2001) The genesis of the Wheatstone Bridge. Engineering Science and Education Journal, 10(1):37–40.
2. Gregory B. A. (1981) An Introduction to Electrical Instrumentation and Measurement Systems, 2nd edition, MacMillan.
3. Kester W., Bryant J., Jung W., Wurcer S., Kitchen C., (1991) Practical design techniques for sensor signal conditioning, Analog Device Inc., ch4.
4. Lotichius J., Wagner S., Kupnik M., Werthschutzky R. (2015) Measurement uncertainty of time-based and voltage-based Wheatstone Bridge readout circuits. IEEE Sensors, 1–4.
5. Fraden J. (2003) Hand book of modern sensors, Physics, Design and Applications. 3rd Edition, New York.
6. Ong G. T., Chan P. K. (2014) A power-aware chopper-stabilized instrumentation amplifier for resistive Wheatstone Bridge sensors. IEEE Transactions on Instrumentation and Measurement, 63(9):2253–2263.

7. Ghosh S., Mukherjee A., Sahoo K., Sen S. K., Sarkar A. (2015) A novel sensitivity enhancement technique employing Wheatstone's Bridge for strain and temperature measurement. Proceedings of the 2015 Third International Conference on Computer, Communication, Control and Information Technology (C3IT), 2015.
8. Boujamaa E. M., Soulie Y., Mailly F., Latorre L., Nouet P. (2008) Rejection of power supply noise in wheatstone bridges: Application to piezoresistive MEMS. Symposium on Design, Test, Integration and Packaging of MEMS/MOEMS, 2008.
9. Mantenuto P., Ferri G., Marcellis A. De (2014) Uncalibrated automatic bridge-based CMOS integrated interfaces for wide-range resistive sensors portable applications. Microelectronics Journal, 45:589–596.
10. Marcellis A. De, Ferri G., Mantenuto P. (2013) Analog Wheatstone Bridge-based automatie interface for grounded and floating wide range resistive sensors. Sensors and Actuators B: Chemical, 187:371–378.
11. Mantenuto P., Marcellis A. De and Ferri G. (2012) Uncalibrated analog bridge based interface for wide-range resistive sensor estimation. IEEE Sensors Journal, 12(5):1413–1414.
12. Lopez-Martin A. J., Zuza M., Carlosena A. (2002) A CMOS interface for resistive bridge transducers. IEEE International Symposium on Circuits and Systems (ISCAS), 2002.
13. Fauzi N. I. M., Anuar N. F., Hana Herman S., Abdullah W. F. H. (2015) Integrated readout circuit using active bridge for resistive-based sensing. Proceedings Computer Science, 2015.
14. Morgenshtein A., Sudakov-Boreysha L., Dinnar U., Jakobson C. G., Nemirovsky Y. (2004) Wheatstone-bridge readout interface for ISFET/REFET applications. Sensors and Actuators, 18–27.
15. Boujamaa E. M. et al. (2001) A low power interface circuit for resistive sensors with digital offset compensation. IEEE International Symposium on Circuits and Systems, 2001.
16. Stefanescu D. (2011) Strain gauges and wheatstone bridges-basic instrumentation and new applications for electrical measurement of non-electrical quantities. Proc. 8th Int. Multi-Conf. SSD, 2011.
17. Lopez-Martin A. J., Osa J. I., Zuza M., Carlosena A. (2003) Analysis of a negative impedance converter as a temperature compensator for bridge sensors. IEEE Transactions on Instrumentation and Measurement, 52(4):1068–1072.
18. Kopsytynski P., Obermeier E. (1989) An interchangeable silicon pressure sensor with on-chip compensation circuitry. Sensors and Actuators, 18(3):239–245.
19. Johnson C. D., Chen C. (1990) Bridge-to-computer data acquisition system with feedback nulling. IEEE Transactions on Instrumentation and Measurement, 39(3):531–534.
20. Graaf G. De, Wolffenbuttel R.F. (2006) Systematic approach for the linearization and readout of non-symmetric impedance bridges. IEEE Transactions on Instrumentation and Measurement, 55(5):1566–1572.
21. Madhu N. M., Geetha T., Sankaran P., Jagadeesh V. K. (2017) Linearization of the output of a wheatstone bridge for single active sensor. IEEE Sensors Journal, 17:1696–1705.
22. Maxim Corporation – Application Note AN3450, Positive Analog Feedback compensates Pt100 Transducer, available online at http://pdfserv.maxim-ic.com/en/an/AN3450.pdf (last accessed June 28, 2008).
23. Bacharowski W. (2008) A precision interface for a Resistance Temperature Detector (RTD). National Semiconductor Corporation, 2008 – available online at http://www.national.com/nationaledge/dec04/article.html (last accessed June 28, 2008).
24. Azhari S. J., Kaabi H. (2000) AZKA cell, the current-mode alternative of wheatstone bridge. IEEE Transactions on Circuits and Systems I, 47(9):1277–1284.
25. Ghallab Y. H., Badawy W. (2006) A New topology for a current-mode wheatstone bridge. IEEE Transactions on Circuits and Systems II, Express Briefs, 53(1):18–22.
26. Khan A. A., Al-Turaigi M. A., El-Ela M. A. (1994) Operational floating current conveyor: Characteristics, Modeling and Applications. IEEE Instrumentation and Measurement Technology Conference, 2:788–791, 1994.
27. Director S. W., Rohrere R. A. (1969)The generalized adjoint network and network sensitivities. IEEE Transactions on Circuits Theory, CT-16:318–323.

28. Roberts G. W., Sedra A. S. (1989) All current-mode frequency selective circuits. Electronics Letters, 25(12):759–761.
29. Mucha I. (1995) Current operational amplifiers: basic architecture, properties, exploitation and future. Analog Integrated Circuits and Signal Processing, 7(3):243–255.
30. Safari L., Minaei S. (2017) A novel COA-based electronically adjustable current-mode instrumentation amplifier topology. International Journal of Electronics and Communications 82:285–293.
31. Farshidi E. (2008) Simple realization of CMOS current-mode wheatstone bridge. IEEE Signals Circuits and Systems International Conference, 2008.
32. Tanaphatsiri C., Jaikla W., Siripruchyanun M. (2008) A current-mode wheatstone bridge employing only single DO-CDTA.IEEE Asia Pacific Conference on Circuits and Systems, 2008.
33. Jaikla W., Siripruchyanun M. (2006) New low temperature-sensitive and electronically controllable configurations for the measurement of small resistance changes. Proceedings of the international technical conference on circuits/systems, computers and communications, 2006.
34. Barthélemy H., Kussener E., Meillère S. (2010) CMOS instrumentation-amplifier based on ASKA cell. Proceedings of the 8th IEEE International NEWCAS Conference, 2010.
35. Safari L., Barile G., Ferri G., Stornelli V. (2018) New resistor free current mode wheatstone bridge topologies with intrinsic linearity. IEEE Prime Conference, 2018.
36. Gift S., Maundy B. (2006) New configurations for the measurement of small resistance changes. IEEE Transaction on Circuits and Systems II, 53(3):178–182.

Chapter 4
Current-Mode Instrumentation Amplifiers Using Current Conveyors

4.1 CCII-Based Basic CMIAs

4.1.1 Wilson Current-Mode Instrumentation Amplifier

4.1.1.1 Basics of Operation

The first current-mode instrumentation amplifier (CMIA) using current conveyors as active building block was introduced by Wilson in 1998 [1]. In the Wilson CMIA illustrated in Fig. 4.1, two plus type second-generation current conveyors ($CCII^+$s) are employed. Input voltages V_1 and V_2 are applied to Y terminals of $CCII^+_1$ and $CCII^+_2$, respectively, which have ideally infinite impedances. By assuming ideal current conveyors, the input voltages at Y terminals are transferred to X terminals with unity gain. Consequently, a current I is produced across R_G equal to:

$$I = \frac{V_1 - V_2}{R_G} \tag{4.1}$$

Assuming that the current conveyors are ideal, current I is transferred to both output terminals through conveyor action between X and Z terminals and is converted into a proportional voltage by load resistors R_1 and R_2 as:

$$Vo_1 = \frac{R_1}{R_G}\left(V_1 - V_2\right) \tag{4.2}$$

$$Vo_2 = \frac{R_2}{R_G}\left(V_1 - V_2\right) \tag{4.3}$$

For common-mode inputs of $V_1 = V_2 = V_{cm}$, the voltage at both terminals of R_G is equal and ideally no current is produced across it. Therefore, the common-mode

G. Ferri et al., *Current-Mode Instrumentation Amplifiers*, Analog Circuits and Signal Processing, https://doi.org/10.1007/978-3-030-01343-1_4

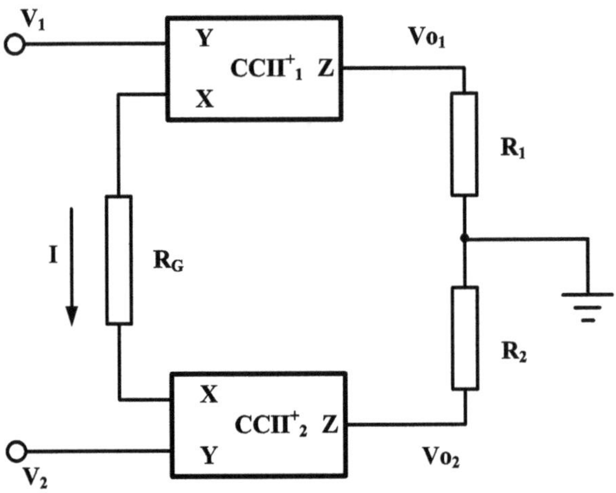

Fig. 4.1 Basic CMIA proposed by Wilson [1]

gain is zero without the need for any resistor matching, whereas differential-mode inputs of $V_1 = -V_2 = V_{dm}/2$ produce a current across R_G and a proportional output voltage. The differential gain can be varied by R_1 and R_2 independently of gain. The interesting feature of Wilson CMIA is that as the differential gain is set by only two resistors and there is no need for a resistor matching to reach high CMRR. The final value of CMRR is determined by the matching between current conveyors and is independent of the voltage gain. In addition, its bandwidth is large because the current conveyors are operating in open loop without the constant gain-bandwidth product restriction which exists in 3-Op-Amp based IA as discussed in Chap. 1. Due to these features, this circuit is suitable for wide-bandwidth applications.

4.1.1.2 Effect of Current Conveyors Non-idealities

Figure 4.2 shows the Wilson CMIA by considering current conveyors non- idealities. Parasitic parameters C_{xi}, r_{xi}, C_{zi} and r_{zi} (for $i = 1, 2$) are the capacitance of X terminal, resistance at X terminal, capacitance at Z terminal and resistance at Z terminal of the associated current conveyor, respectively.

Using the analysis performed in [2] and by assuming a non unity voltage and current transfer gains and parasitic impedances at X terminals of current conveyors, the transfer functions will be:

$$Vo_1 = \frac{\beta_1 R_1}{\left(R_G + rx_1 + rx_2\right)\left(1 + \frac{S}{\omega_{Z1}}\right)}\left(\alpha_1 V_1 - \alpha_2 V_2\right) + \frac{SC_{x1} R_1 \alpha_1 \beta_1}{\left(1 + \frac{S}{\omega_{Z1}}\right)} V_1 \tag{4.4}$$

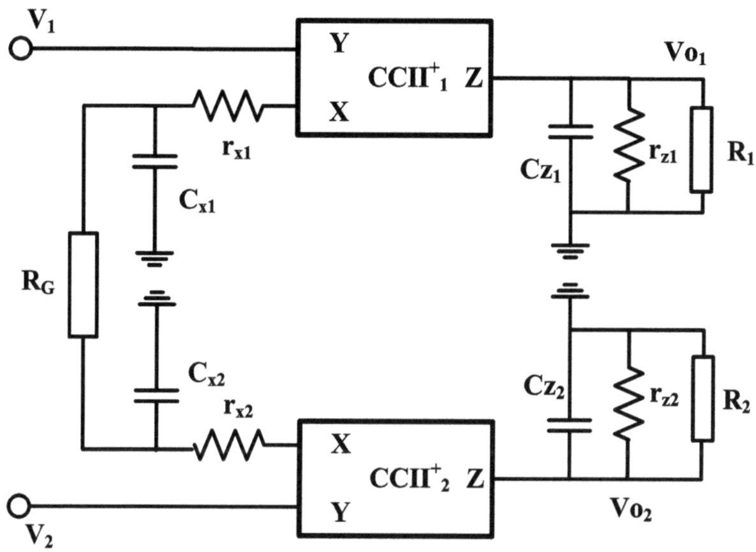

Fig. 4.2 Wilson CMIA by considering conveyors non-idealities

$$Vo_2 = -\frac{\beta_1 R_2}{(R_G + rx_1 + rx_2)\left(1+\frac{S}{\omega_{Z2}}\right)}(\alpha_1 V_1 - \alpha_2 V_2) + \frac{SC_{x2}R_2\alpha_2\beta_2}{\left(1+\frac{S}{\omega_{Z2}}\right)}V_2 \tag{4.5}$$

where α_i and β_i (for $i = 1,2$) are the voltage gain between Y and X terminals and current gain between X and Z terminals of the related CCII respectively. Here, by assuming $R_1, R_2 \ll r_{z1}, r_{z2}$, it is: $\omega_{z1} = 1/(R_1 C_{Z1})$ and $\omega_{z2} = 1/(R_2 C_{Z2})$.

For common-mode inputs of $V_1 = V_2 = V_{cm}$, the common-mode gain is:

$$A_{c1} = \frac{V_{oc1}}{V_{cm}} = \frac{\beta_1 R_1}{(R_G + rx_1 + rx_2)\left(1+\frac{S}{\omega_{Z1}}\right)}(\alpha_1 - \alpha_2) + \frac{SC_{x1}R_1\alpha_1\beta_1}{\left(1+\frac{S}{\omega_{Z1}}\right)} \tag{4.6}$$

$$A_{c2} = \frac{V_{oc2}}{V_{cm}} = -\frac{\beta_2 R_2}{(R_G + rx_1 + rx_2)\left(1+\frac{S}{\omega_{Z2}}\right)}(\alpha_1 - \alpha_2) + \frac{SC_{x2}R_2\alpha_2\beta_2}{\left(1+\frac{S}{\omega_{Z2}}\right)} \tag{4.7}$$

According to Eq. (4.6) and Eq. (4.7), in real case common-mode gain is non-zero and determined by matching between two CCIIs. The interesting point is that due to non-zero value of C_x, even for matched conveyors with $\alpha_1 = \alpha_2$, the common-mode gain is not zero.

Differential-mode gains (for differential-mode inputs of $V_1 = -V_2 = V_{dm}/2$) will be:

$$A_{d1} = \frac{V_{od1}}{V_{dm}} = \frac{\beta_1 R_1}{2\left(1+\dfrac{S}{\omega_{Z1}}\right)}(\alpha_1+\alpha_2)+\frac{SC_{x1}R_1\alpha_1\beta_1}{\left(1+\dfrac{S}{\omega_{Z1}}\right)} \tag{4.8}$$

$$A_{d2} = \frac{V_{od2}}{V_{dm}} = -\frac{\beta_2 R_2}{2\left(R_G+rx_1+rx_2\right)\left(1+\dfrac{S}{\omega_{Z2}}\right)}(\alpha_1+\alpha_2)-\frac{SC_{x2}R_2\alpha_2\beta_2}{\left(1+\dfrac{S}{\omega_{Z2}}\right)} \tag{4.9}$$

As it is seen from Eq. (4.8) and Eq. (4.9), Wilson CMIA suffers from gain error and limited accuracy (due to non-zero value of r_{x1} and r_{x2}). It is also worth mentioning that to preserve circuit bandwidth and gain, it is necessary to include voltage buffers between output terminals of current conveyors and load. The latter also provides low output impedance and high drive capability for instrumentation amplifier. In the case where only one output is required, the other output terminal can be connected directly to ground.

4.1.2 Gift Current-Mode Instrumentation Amplifier

In [3], Gift introduced a new version of Wilson basic CMIA where the problems of the limited accuracy and gain error were effectively resolved. In Gift enhanced CMIA, Operational Conveyors (OCs) were used instead of current conveyors. Figure 4.3 shows an OC which consists of an Op-Amp, working in conjunction with a CCII. To eliminate the effect of r_x, the input of current conveyor is inside the feedback loop of the Op-Amp. The schematic of the Gift enhanced CMIA is shown in Fig. 4.4 and its low-frequency transfer function, for differential inputs and with the same significance of symbols, is found as [3]:

$$A_d = \frac{kR_1}{2r_x+(1+k)R_G} \tag{4.10}$$

where k is Op-Amp gain. For $k \gg 1$, Eq. (4.10) can be expressed as:

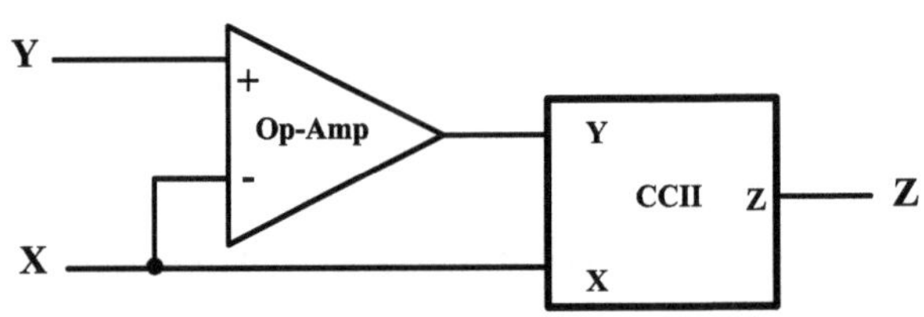

Fig. 4.3 Operational conveyor schematic [3]

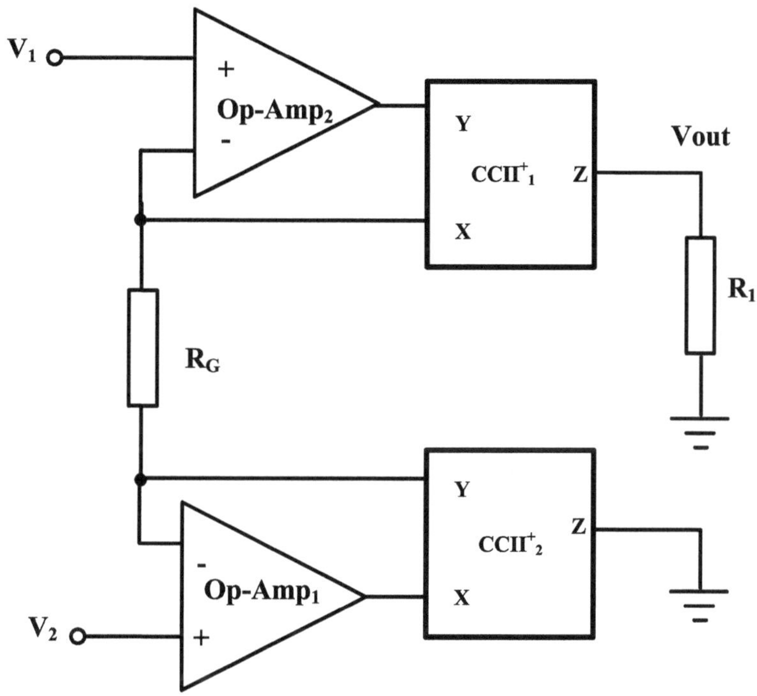

Fig. 4.4 Gift Basic CMIA [3]

$$A_d \cong \frac{kR_1}{(1+k)R_G} \cong \frac{R_1}{R_G} \tag{4.11}$$

As it is seen from Eq. (4.11), the transfer function depends only on external resistor and the effect of r_x is effectively eliminated from the transfer function.

4.2 Current Conveyor-Based Enhanced CMIAs with Double Common-Mode Cancellation

4.2.1 Khan et al. Current-Mode Instrumentation Amplifier

In [2], Khan et al. introduced the improved version of Wilson CMIA in which two effective modifications are applied:

1. the output of second conveyor was inverted by means of a third $CCII^+$;
2. the resulting output current has been added to that of the first $CCII^+$.

By these modifications, a current subtraction node is made at output which is used to further cancel the remaining common-mode output from first stage.

The circuit is shown in Fig. 4.5. Compared to Wilson CMIA of Fig. 4.1, the differential gain is increased by a factor of two. For matched CCIIs, the common-mode gain also approaches a very low value because the common-mode currents produced by parasitics at X terminals are subtracted at output node. Like Wilson CMIA, the differential gain is inversely proportional to r_x and the accuracy is limited by r_x tolerance. As a further point, for practical use, a voltage buffer is necessary at output node.

4.2.2 Su and Lidgey Current-Mode Instrumentation Amplifier

Another CMIA has been proposed by Su and Lidgey [4] and is shown in Fig. 4.6. It is formed by adding an Op-Amp based differential amplifier to the output of Wilson basic CMIA. In it, the outputs of CCIIs is subtracted by a differential amplifier made of an Op-Amp and two resistors. Unfortunately, in the circuit, the CMRR is limited because current transfer errors of current conveyors are different significantly from each other. The reason is that their output terminals are connected to different impedances. However, the circuit has high input (ideally infinity) and low output (ideally zero) impedances which makes it attractive for cascading.

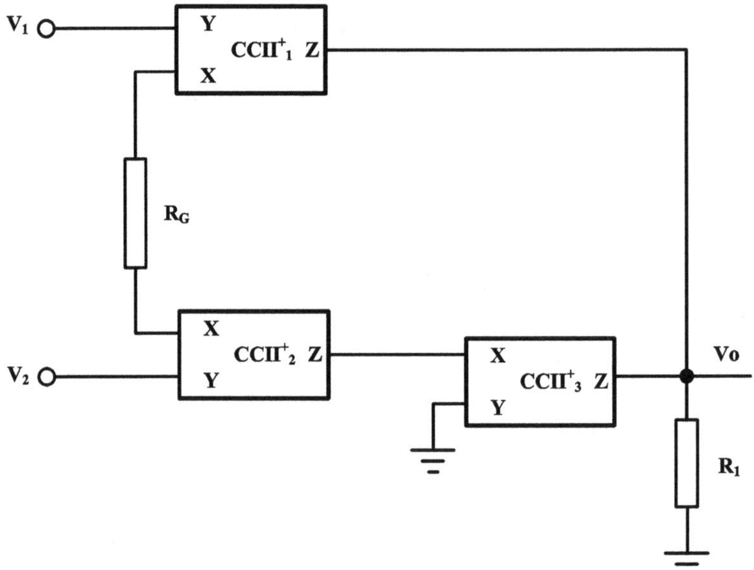

Fig. 4.5 CMIA proposed by Khan et al. [2]

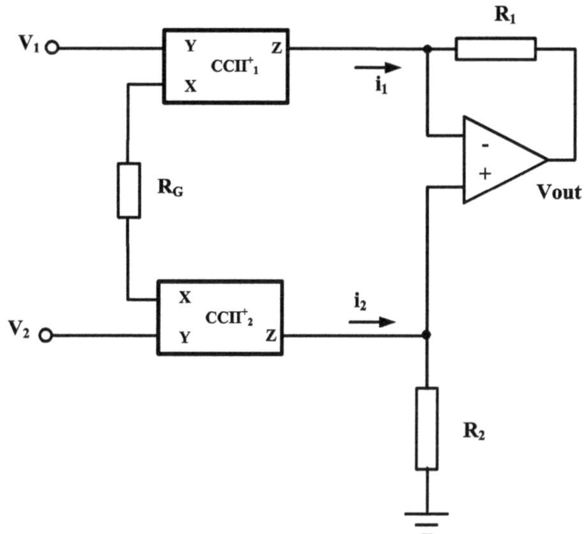

Fig. 4.6 CMIA proposed by Su and Lidgey [4]

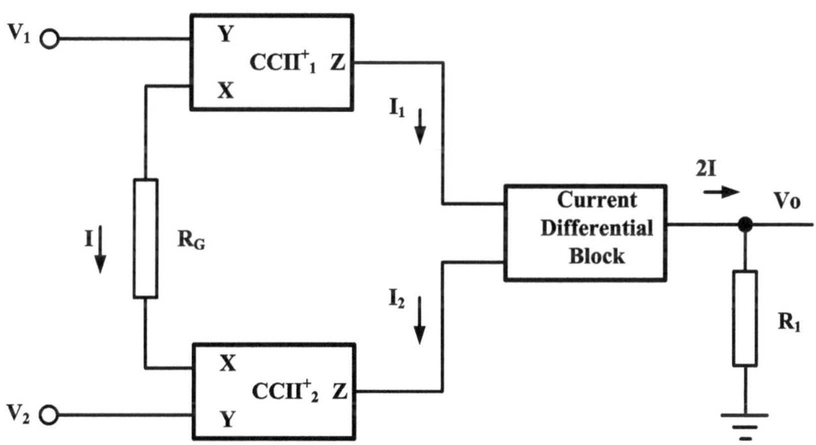

Fig. 4.7 CMIA proposed by Galanis [5]

4.2.3 Galanis Current-Mode Instrumentation Amplifier

The CMIA topology proposed by Galanis et al. [5] is shown in Fig. 4.7. This circuit is formed by connecting a current differencing block made of a single current mirror to the output of Wilson basic CMIA. Similar to Khan CMIA, the output currents of input CCIIs are subtracted at output nodes. However, here the subtraction is performed by means of a current differential block. This topology yields higher differential-mode gain that is two times larger than Wilson CMIA and improved CMRR. The disadvantage of this circuit is that its output impedance depends on R_1, thus a voltage buffer may be needed before connecting to the load.

4.2.4 Gkotsis Current-Mode Instrumentation Amplifier

The CMIA introduced by Gkotsis et al. [6] is shown in Fig. 4.8. Similar to the circuit proposed by Galanis et al. in [5], a current differencing block is used to produce the difference of output currents of current conveyors. However, compared to the Galanis CMIA, in [6], the current differencing block is formed by three current mirrors. In addition, it includes a voltage buffer and an offset cancellation circuit which makes it more suitable for practical use.

4.2.5 Koli Current-Mode Instrumentation Amplifier

In [7] Koli et al. employed a composite current conveyor (a high performance block realized by two or more current conveyors) for CMIA implementation. Figure 4.9 shows a composite current conveyor constructed by two $CCII^+$s. In this conveyor, the negative feedback loop reduces impedance at X terminal and enhances the voltage follower accuracy between Y and X terminals. Resistor R_F is used to stabilize the feedback loop in bipolar realization of the conveyors.

The CMIA topology in which composite conveyors were employed is shown in Fig. 4.10. Here, $CCII_3$ is used to invert the output current of $CCII_2$. The Z terminal of $CCII_3$ is connected, through R_3-C_1, to the Z terminal of $CCII_1$ to make a current

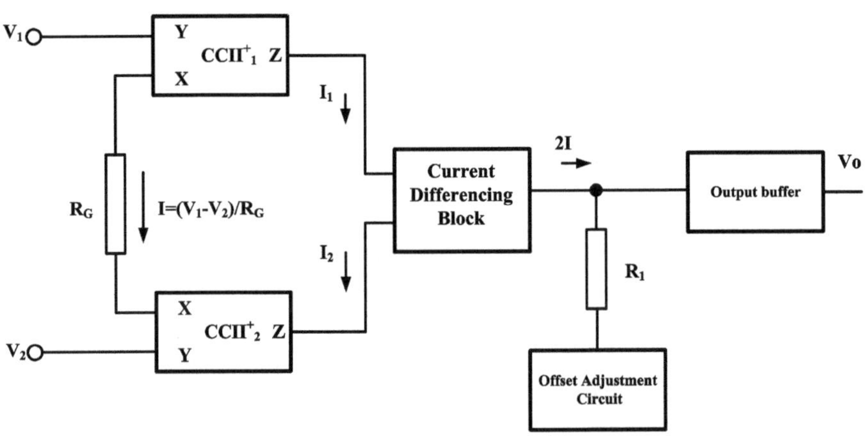

Fig. 4.8 CMIA proposed by Goktis [6]

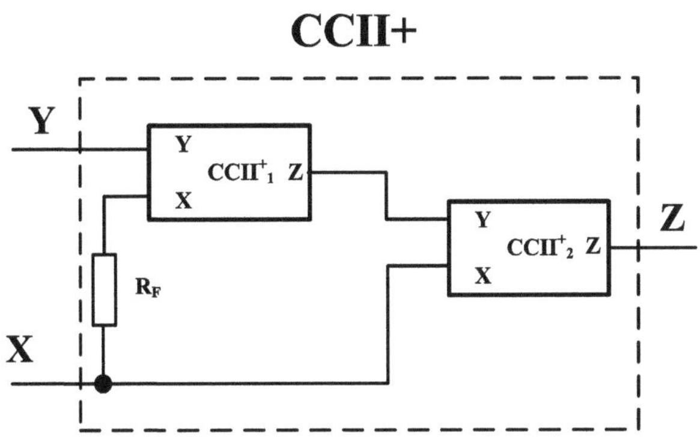

Fig. 4.9 Composite current conveyor using two current conveyors [7]

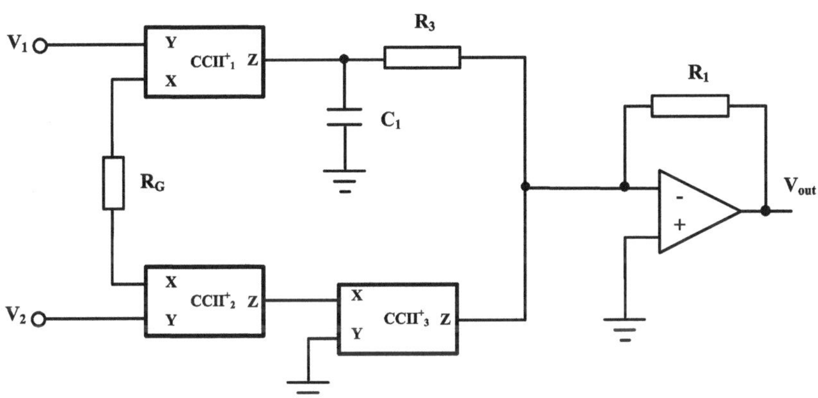

Fig. 4.10 CMIA proposed by Koli [7]

subtraction node. The amplifier made of Op-Amp and resistor R_1 is used to achieve the required gain. The common-mode performance is usually only slightly improved due to the reduced voltage follower gain error. In this circuit the CMRR sensitivity to current conveyors gain mismatch is reduced. In addition, as the composite conveyor has lower r_x, the gain error is also reduced.

In Fig. 4.10 the CCII+ can be replaced by a composite CCII, whose symbol is CCII+.

4.2.6 Azhari and Fazlalipour Current-Mode Instrumentation Amplifier

The basic structure of the CMIA topology proposed in [8] is shown in Fig. 4.11. It consists of two CCIIs, a grounded load (R_1) and a gain controlling resistor (R_G). The output of unloaded CCII is connected to the input X of first CCII. This connection

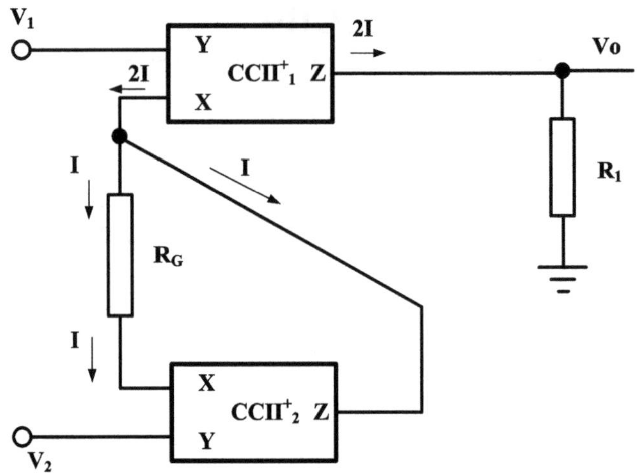

Fig. 4.11 CMIA proposed by Azhari and Fazlalipour [8]

allows the further subtraction of common-mode currents and the addition of differential-mode currents right at the input of first CCII without the need of an extra current differential block. This connection allows the differential-mode gain becomes two times greater than circuits reported by Wilson [1] and Gift [3]. This method is also referred to as error correction technique.

4.2.7 Gift et al. Enhanced Current-Mode Instrumentation Amplifier

In [9] Gift et al. introduced a new CMIA in which the error correction technique is applied to their previous enhanced CMIA reported in [8]. The circuit which is shown in Fig. 4.12 is a high-performance CMIA with high accuracy and high CMRR. This configuration exhibits improved performance compared to CMIA reported in [8] because the voltage transfer error and r_x of operational conveyor is significantly reduced. This circuit has a more accurate transfer function than any of the other topologies described. However, like the Galanis circuit of Fig. 4.7, the output impedance depends on R_1, so a voltage buffer at the output may be needed.

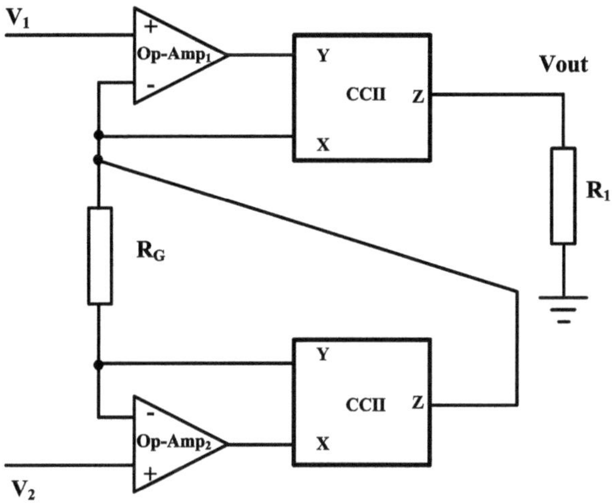

Fig. 4.12 Enhanced CMIA proposed by Gift et al. [9]

References

1. Wilson B.(1989) Universal conveyor instrumentation amplifier.Electronics Letters, 25:470–471.
2. Gift S. J. G. (2001) An enhanced current-mode instrumentation amplifier.IEEE Transactions on Instrumentation and Measurement, 50(1):85–88.
3. Khan A. A., Al-Turaigi M.A., El-Ela M. A. (1995) An improved current-mode instrumentation amplifier with bandwidth independent of gain. IEEE Transactions on Instrumentation and Measurement, 44:887–891.
4. Su W., Lidgey F. G. (1995) Common-mode rejection in current-mode instrumentation amplifiers. Analog Integrated Circuits and Signal Processing, 7(3):257–260.
5. Galanis C., Haritantis I. (1996) An improved current-mode instrumentation amplifier. Proceedings of the third IEEE International Conference on Electronics, Circuits and Systems, 1996.
6. Gkotsis I., Souliotis G., Haritantis I. (1998) Instrumentation amplifier based analogue interface. IEEE International Conference on Electronics, Circuits and Systems, 317–320, 1998.
7. Koli K. A. I., Halonen A. I. (2000) CMRR enhancement techniques for current mode instrumentation amplifiers. IEEE Transactions on Circuit and Systems-I,622–632.
8. Fazlalipoor H., Azhari S. J. (2000) A novel current mode instrumentation amplifier (CMIA) topology. IEEE Transactions on Instrumentation and Measurement, 49(6):1272–1277.
9. Gift S. J. G., Maundy B., Muddeen F. (2007) High-performance current-mode instrumentation amplifier circuit. International Journal of Electronics, 94(11):1015–1024.

Chapter 5
Current-Mode Instrumentation Amplifiers Based on Various Current-Mode Building Blocks

5.1 Current Input-Current Output (I-I) CMIAs

In Chap. 3, two I-I CMIA topologies based on Current-Mode Operational Amplifier (COA) and Operational Floating Conveyor (OFCC) are discussed. In this section, three other topologies based on Current Differencing Buffered Amplifier (CDBA), Current Feedback Operational Amplifier (CFOAs) and OFCC are studied.

5.1.1 CDBA-Based I-I CMIA

In [1], a CMIA-based on CDBA is reported. Figure 5.1 shows the circuit symbol of CDBA which is a four terminal current-mode building block where *p* and *n* are low impedance current input terminals, Z is the high impedance current output terminal and W is the low impedance voltage output terminal. There is a voltage tracking between W and Z terminals and the relations between terminal voltages and signals are expressed as [1, 2]:

$$I_z = \beta_p I_p - \beta_n I_n, \quad V_W = \alpha V_z, V_n = r_n I_n, \ V_p = r_p I_p \tag{5.1}$$

where β_p and β_n are the current gains of *p* and *n* terminals respectively and α is the voltage gain between Z and W terminals. Parasitic resistances at *n* and *p* terminals are shown as r_n and r_p, respectively. The impedance at Z terminal, r_z must be very large. In the ideal case, $\beta_p = \beta_n = 1$, $\alpha = 1$, $r_z = \infty$ and $r_n = r_p = 0$.

The CDBA-based I-I CMIA of [1] is shown in Fig. 5.2 and consists of two CDBAs and two resistors.

In this circuit, two input currents (I_{in1}, I_{in2}) are applied to low impedance *p* and *n* terminals of $CDBA_1$. The difference between I_{in1} and I_{in2} are produced at Z terminal and, by assuming $r_z \gg R_1$, is converted to a voltage equal to $V_{Z1} \approx (\beta_{p1} I_{in1} - \beta_{n1} I_{in2}) R_1$.

G. Ferri et al., *Current-Mode Instrumentation Amplifiers*, Analog Circuits and Signal Processing, https://doi.org/10.1007/978-3-030-01343-1_5

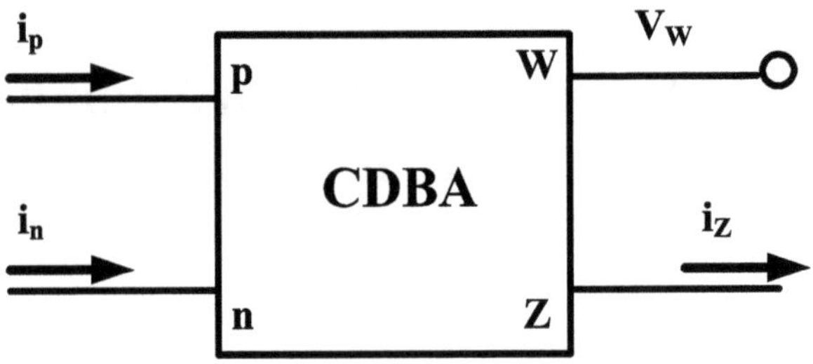

Fig. 5.1 Schematic symbol of CDBA [1]

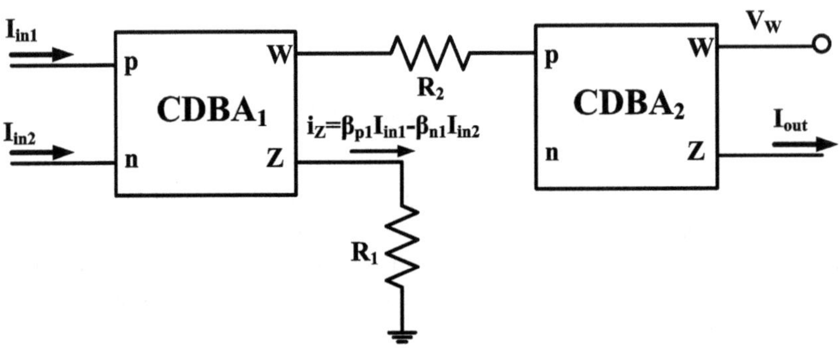

Fig. 5.2 CDBA based I-I CMIA of [1]

The high impedance at Z terminal of CDBA allows direct connection of resistor R_1 to Z_1 terminal. The voltage produced at Z terminal of $CDBA_1$ is copied to its W terminal. By assuming the voltage at p terminal of $CDBA_2$ at ground (due to very low value of r_p), the voltage across R_2 is equal to V_z. Therefore, a current equal to V_z/R_2 is injected into p terminal of $CDBA_2$ which is conveyed into its Z terminal by a gain of β_{p2} producing I_{out} as:

$$I_{out} = \alpha_1 \beta_{p2} \frac{R_1}{R_2 + r_{p2} + r_{w1}} \left(\beta_{p1} I_{in1} - \beta_{n1} I_{in2} \right) \tag{5.2}$$

For common-mode input currents ($I_{in1} = I_{in2} = I_{cm}$), using Eq. (5.2), the common-mode gain is:

$$A_{cm} = \frac{I_{out}}{I_{cm}} = \alpha_1 \beta_{p2} \frac{R_1}{R_2 + r_{p2} + r_{w1}} \left(\beta_{p1} - \beta_{n1} \right) \tag{5.3}$$

For differential-mode input currents ($I_{in1} = -I_{in2} = I_{dm}/2$), using Eq. (5.2) the differential-mode gain is:

$$A_{dm} = \frac{I_{out}}{I_{dm}} = \frac{\alpha_1 \beta_{p2} R_1}{2(R_2 + r_{p2} + r_{w1})} \left(\beta_{p1} + \beta_{n1} \right) \tag{5.4}$$

From Eqs. (5.3) and (5.4), CMRR is found as:

$$CMRR = \frac{A_{dm}}{A_{cm}} = \frac{\beta_{p1} + \beta_{n1}}{2\left(\beta_{p1} - \beta_{n1}\right)} \tag{5.5}$$

As Eq. (5.5) indicates, the CMRR of the CMIA of Fig. 5.2 depends only on the matching between β_{p1} and β_{n1}. From Eq. (5.4), to reduce differential-mode gain error, the condition $R_2 \gg (r_{p2} + r_{w1})$ must be satisfied. Differential-mode gain can be easily varied by changing either R_2 or R_1.

5.1.2 CFOA Based I-I CMIA

In [3] a I-I CMIA based on CFOA is reported. Figure 5.3 shows the symbol of the CFOA. It is a four terminal device with high impedance voltage input Y terminal, low impedance current input X terminal, high impedance current output Z terminal and low impedance voltage output W terminal. The relations between voltage and current signals are expressed as:

$$I_z = \beta I_x, \quad V_x = \alpha_1 V_Y, \quad V_W = \alpha_2 V_Z, \quad V_x = r_x I_x \tag{5.6}$$

where β is the current gain between X and Z terminals, α_1 is the voltage gain between Y and X terminals and α_2 is the voltage gain between W and Z terminals. The parasitic impedance at X terminal is also shown by r_x.

Figure 5.4 shows the I-I CFOA based CMIA IA of [3] which consists of two CFOAs and two resistors. In this circuit, I_{in1} is directly connected to Z node of

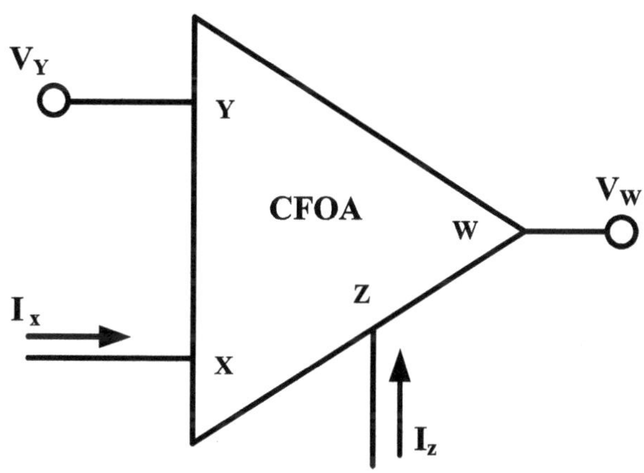

Fig. 5.3 Schematic representation of CFOA [3]

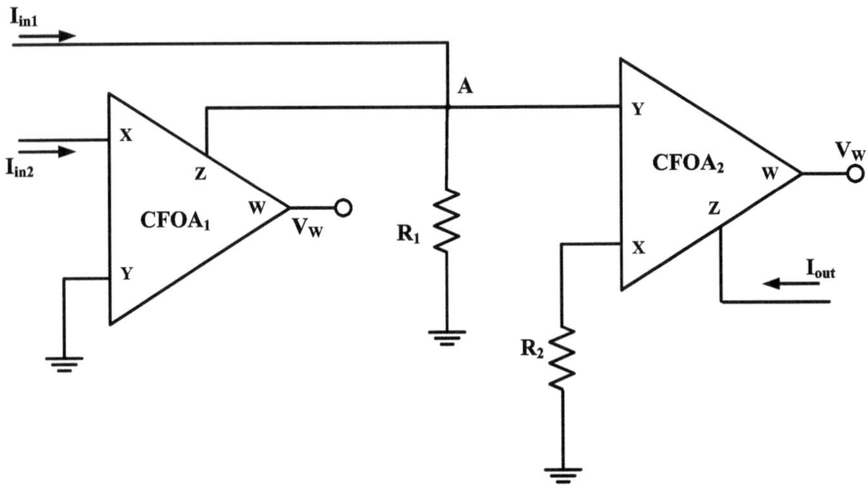

Fig. 5.4 CFOA based I-I CMIA [3]

$CFOA_1$ where it is subtracted from $I_{Z1}(=\beta_1 I_{in2})$ and the resulted output current is converted to a proportional voltage by R_1 at node A:

$$V_A = \left(I_{in1} - \beta_1 I_{in2}\right) R_1 \tag{5.7}$$

The voltage produced at node A is conveyed to X terminal of $CFOA_2$ where it is converted to current by R_2. The current at R_2 is transferred to Z terminal of $CFOA_2$ and by assuming $R_1 \ll r_{z1}, r_{y1}$, the resulted output is:

$$I_{out} = \alpha_2 \left(I_{in1} - \beta_1 I_{in2}\right) \frac{R_1}{R_2 + r_{x2}} \tag{5.8}$$

Using Eq. (5.8), for common-mode input currents of $I_{in1} = I_{in2} = I_{cm}$, common-mode gain is:

$$A_{cm} = \frac{I_{outc}}{I_{cm}} = \alpha_2 \left(1 - \beta_1\right) \frac{R_1}{R_2 + r_{x2}} \tag{5.9}$$

For differential-mode input currents of $I_{in1} = -I_{in2} = I_{dm}/2$, differential-mode gain is:

$$A_{dm} = \frac{I_{outd}}{I_{dm}} = \frac{\alpha_2 \left(1 + \beta_1\right)}{2} \frac{R_1}{R_2 + r_{x2}} \tag{5.10}$$

Using Eqs. (5.9) and (5.10), CMRR is calculated as:

$$CMRR = \frac{A_{dm}}{A_{cm}} = \frac{1 + \beta_1}{2\left(1 - \beta_1\right)} \tag{5.11}$$

In the CFOA-based CMIA of Fig. 5.4, to achieve high CMRR, β_1 must be as close as possible to unity. Although CMRR does not depend on the matching between $CFOA_1$ and $CFOA_2$, there is an imbalance between input impedances for I_{in1} and I_{in2} which reduces overall CMRR especially if the output impedance of I_{in1} is not much larger than the equivalent impedance at node A. In this circuit, differential-mode gain can be adjusted by R_1 or R_2. To reduce gain error, the condition $R_2 \gg r_{x2}$ must be satisfied.

5.1.3 OFCC-Based I-I CMIA

In [4] a I-I CMIA topology based on OFCC as basic building block is presented. Figure 5.5 shows the symbolic representation of OFCC, which is a current-mode five terminal basic building block. X is a current input low impedance port, Y is a voltage input high impedance port, W is a voltage output low impedance port, while Z^+ and Z^- are current output high impedance ports. The voltage at Y port is conveyed to X port and the current of port W is transferred to Z^+ and Z^- ports in phase and out-of-phase, respectively. The input current at port X is multiplied to OFCC open loop trans-impedance gain Z_t and appears at port W making OFCC suitable for closed loop operations. The relation between ports voltages and currents are described in Eq. (5.12) where α is the voltage gain between Y and X terminals, r_x is the parasitic impedance at port X, Z_t is trans-impedance gain, β is current gain between Z^+ and W ports and γ is current gain between Z^- and W ports.

$$V_x = \alpha V_y + I_x r_x, \quad V_W = Z_t I_x, \quad I_{z+} = \beta I_W, \; I_{z-} = \gamma I_W \tag{5.12}$$

The I-I OFCC based CMIA of [4] is shown in Fig. 5.6 where $OFCC_1$ and $OFCC_2$ are performing as closed loop current amplifiers. The input currents I_{in1} and I_{in2} are amplified by $OFCC_1$ and $OFCC_2$ respectively. The Z^+ output of $OFCC_1$ is connected to Z^- of $OFCC_2$ to make a current subtraction node. Therefore the amplified input currents are subtracted and the resulted output is further amplified by $OFCC_3$. A simple analysis gives the output current as:

$$I_{out} = \left\{ \left(1 + \frac{R_2}{R_1}\right) \beta_1 I_{in1} - \left(1 + \frac{R_4}{R_3}\right) \gamma_2 I_{in2} \right\} \left(1 + \frac{R_6}{R_5}\right) \beta_3 \tag{5.13}$$

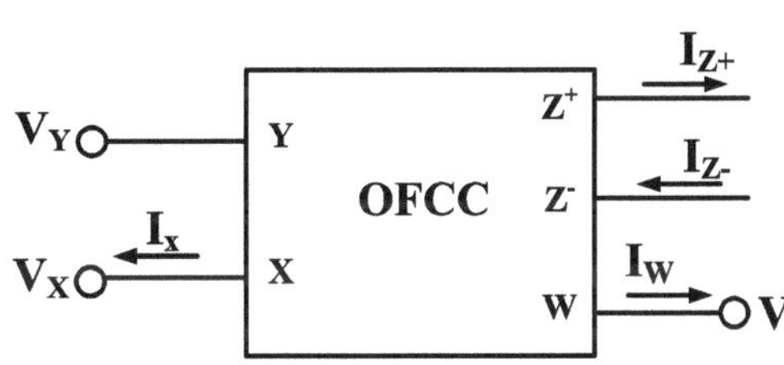

Fig. 5.5 Schematic representation of OFCC [4]

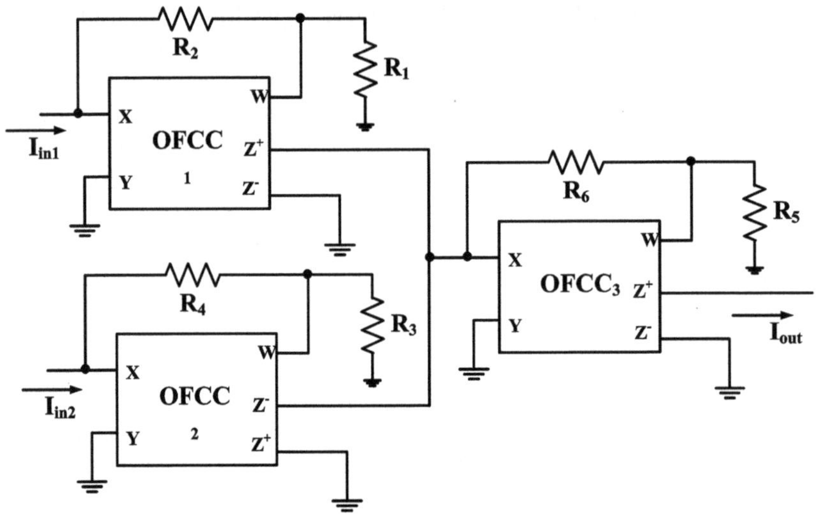

Fig. 5.6 OFCC based I-I CMIA of [4]

For common-mode input currents of $I_{in1} = I_{in2} = I_{cm}$, from Eq. (5.13), common-mode gain is:

$$A_{cm} = \frac{I_{out}}{I_{cm}} = \left\{\left(1+\frac{R_2}{R_1}\right)\beta_1 - \left(1+\frac{R_4}{R_3}\right)\gamma_2\right\}\left(1+\frac{R_6}{R_5}\right)\beta_3 \tag{5.14}$$

For differential-mode input currents of $I_{in1} = -I_{in2} = I_{dm}/2$, from Eq. (5.13), differential-mode gain is:

$$A_{dm} = \frac{I_{out}}{I_{dm}} = \frac{\left\{\left(1+\frac{R_2}{R_1}\right)\beta_1 + \left(1+\frac{R_4}{R_3}\right)\gamma_2\right\}\left(1+\frac{R_6}{R_5}\right)\beta_3}{2} \tag{5.15}$$

From Eqs. (5.14) and (5.15), CMRR is found as:

$$CMRR = \frac{A_{dm}}{A_{cm}} = \frac{\left(1+\frac{R_2}{R_1}\right)\beta_1 + \left(1+\frac{R_4}{R_3}\right)\gamma_2}{2\left[\left(1+\frac{R_2}{R_1}\right)\beta_1 - \left(1+\frac{R_4}{R_3}\right)\gamma_2\right]} \tag{5.16}$$

As Eq. (5.16) states, the CMIA of Fig. 5.6 not only requires resistor matching to achieve high CMRR, but the matching between β_1 and γ_2 is also needed. In addition, although it enjoys closed loop operation which makes it insensitive to parasitic resistance at X terminal, however, large number of used floating and grounded resistors is its main drawback.

5.2 Current Input-Voltage Output (I-V) CMIAs

5.2.1 OFCC-Based I-V CMIA

Figure 5.7 shows the OFCC-based I-V CMIA reported in [4]. Based on the operation of OFCC block given in Sect. 5.1.3, $OFCC_1$ and $OFCC_2$ are used to amplify the input currents I_{in1} and I_{in2} respectively. The Z+ output of $OFCC_1$ and Z- of $OFCC_2$ are connected to form a current subtraction node. Therefore, the outputs produced by $OFCC_1$ and $OFCC_2$ are subtracted and send to $OFCC_3$ which is configured as a trans-impedance amplifier with ideal gain equal to R_5. The relation between output voltage and input currents are:

$$V_{out} = \left\{ \left(1 + \frac{R_2}{R_1}\right)\beta_1 I_{in1} - \left(1 + \frac{R_4}{R_3}\right)\gamma_2 I_{in2} \right\} R_5 \tag{5.17}$$

For common-mode input currents of $I_{in1} = I_{in2} = I_{cm}$, from Eq. (5.17), common-mode gain is:

$$A_{cm} = \frac{V_{out}}{I_{cm}} = \left\{ \left(1 + \frac{R_2}{R_1}\right)\beta_1 - \left(1 + \frac{R_4}{R_3}\right)\gamma_2 \right\} R_5 \tag{5.18}$$

For differential-mode input currents of $I_{in1} = -I_{in2} = I_{dm}/2$, from Eq. (5.17), differential-mode gain is:

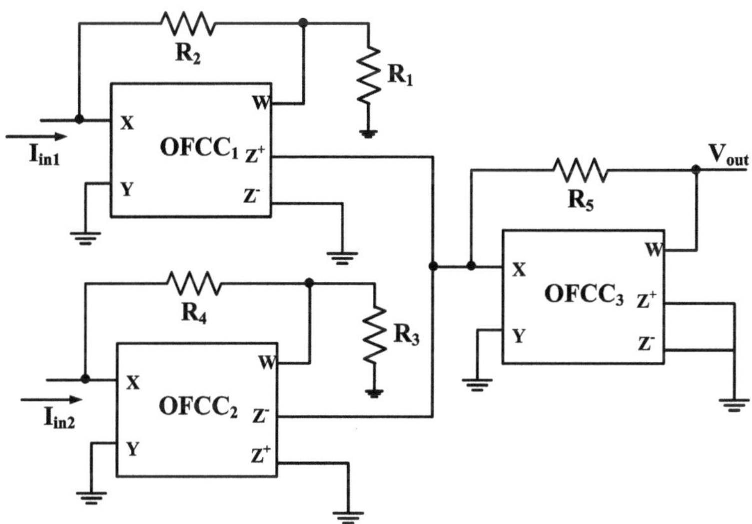

Fig. 5.7 OFCC based I-V CMIA of [4]

$$A_{dm} = \frac{V_{out}}{I_{dm}} = \frac{\left\{\left(1+\frac{R_2}{R_1}\right)\beta_1 + \left(1+\frac{R_4}{R_3}\right)\gamma_2\right\}R_5}{2} \tag{5.19}$$

Based on Eq. (5.19), differential-mode gain can be varied by R_5. From Eqs. (5.18) and (5.19), CMRR is found as:

$$CMRR = \frac{A_{dm}}{A_{cm}} = \frac{\left(1+\frac{R_2}{R_1}\right)\beta_1 + \left(1+\frac{R_4}{R_3}\right)\gamma_2}{2\left[\left(1+\frac{R_2}{R_1}\right)\beta_1 - \left(1+\frac{R_4}{R_3}\right)\gamma_2\right]} \tag{5.20}$$

In this case, high CMRR requires good matching between both the resistors and β_1 and γ_2 parameters. The CMIA of Fig. 5.7 enjoys low impedances at input and output ports. However, large number of resistors and strict matching requirement between them are its main weaknesses.

In [5] an improved version of the OFCC based CMIA of Fig. 5.7 is introduced. This circuit, shown in Fig. 5.8, has a reduced number of resistors. Compared to CMIA of Fig. 5.7 which requires matching between four resistors to achieve high CMRR, in this CMIA matching between only two resistors of R_1 and R_2 is required.

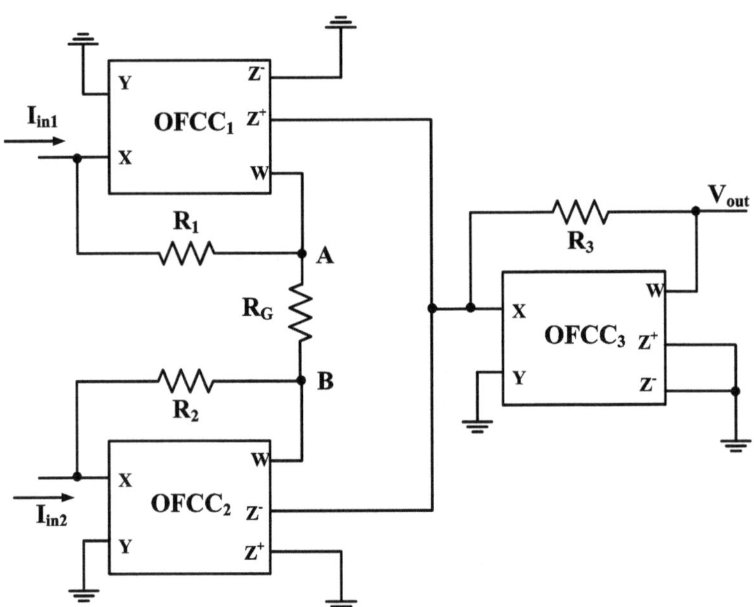

Fig. 5.8 OFCC based I-V CMIA [5]

In this case, differential-mode currents of $I_{in1} = -I_{in2} = I_{dm}/2$ produce two equals in magnitude but opposite polarity voltages at nodes A and B i.e. $V_A = -V_B$. Therefore, for differential-mode input currents, the CMIA of Fig. 5.8 can be shown as Fig. 5.9a. By assuming a high impedance gain, differential-mode gain (for $I_{in1} = -I_{in2} = I_{dm}/2$) is expressed as:

$$A_{dm} = \frac{V_{out}}{I_{dm}} = \frac{\left\{\left(1+\frac{2R_1}{R_G}\right)(\beta_1+\gamma_2)\right\}R_3}{2} \tag{5.21}$$

For common-mode inputs ($I_{in1} = I_{in2} = I_{cm}$), assuming $R_1 = R_2 = R$, nodes A and B have equal voltages. Therefore, as it is shown in Fig. 5.9b, no current flows into R_G, which can be considered as an open circuit. Using Fig. 5.9b, common-mode gain is calculated as:

$$A_{cm} = \{\beta_1 - \gamma_2\}R_3 \tag{5.22}$$

Using Eqs. (5.21) and (5.22), CMRR is found as:

$$CMRR = \frac{1}{2}\left(1+\frac{2R_1}{R_G}\right)\frac{\beta_1+\gamma_2}{\beta_1-\gamma_2} \tag{5.23}$$

This circuit shows low input and output impedances. For high CMRR, matching between β_1 and γ_2 is required. In addition, mismatch between R_1 and R_2 reduces the CMRR value.

5.2.2 *OTRA-Based I-V CMIA*

Operational Trans-Resistance Amplifier (OTRA) is a three terminal current-mode building block in which inputs are currents and the output is of voltage kind [6]. OTRA symbol is shown in Fig. 5.10 and the relations between terminal currents and voltages are expressed in Eq. (5.24) where R_m is trans-resistance gain and r_n and r_p are parasitic resistances of n and p terminals, respectively.

$$V_{out} = R_m\left(I_p - I_n\right), \quad V_n = r_n I_n, \quad V_p = r_p I_p \tag{5.24}$$

Figure 5.11 shows the I-V CMIA based on OTRAs reported in [6]. It comprises three OTRAs and 5 resistors. The input currents are transferred to voltage by $OTRA_1$ and $OTRA_2$. The outputs of first stage can be expressed as:

$$V_{o1} = I_{in1}\left(R_1 + r_{n1}\right) \tag{5.25}$$

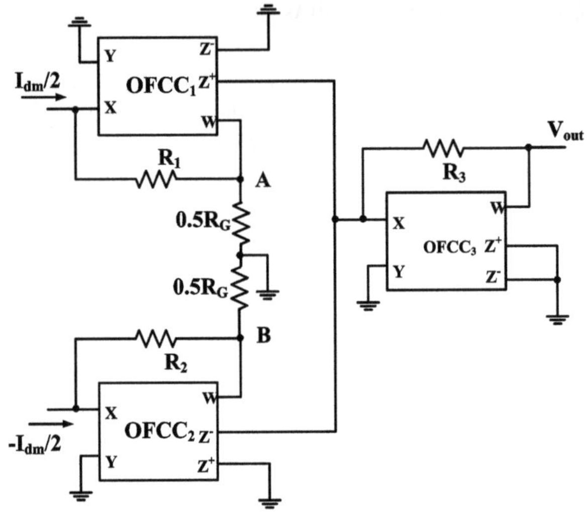

(a)

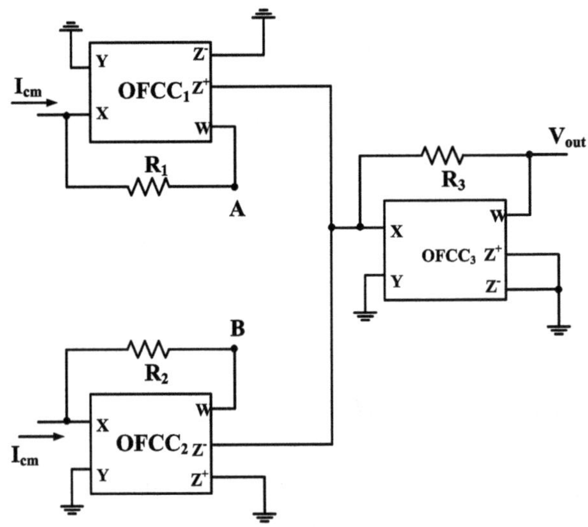

(b)

Fig. 5.9 OFCC based I-V CMIA of [5] in (**a**) differential-mode (**b**) common-mode by assuming $R_1 = R_2$

Fig. 5.10 Schematic representation of OTRA [6]

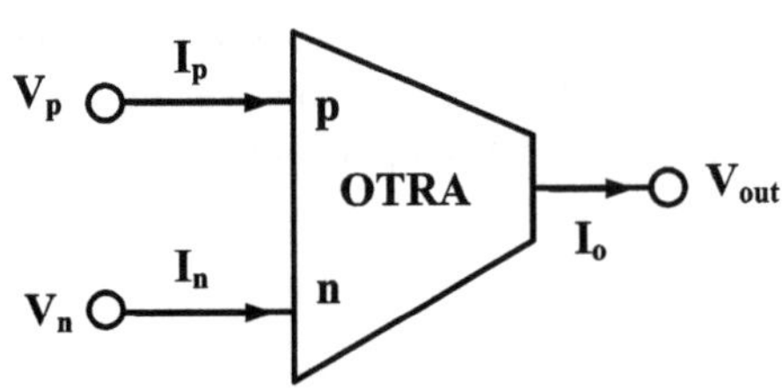

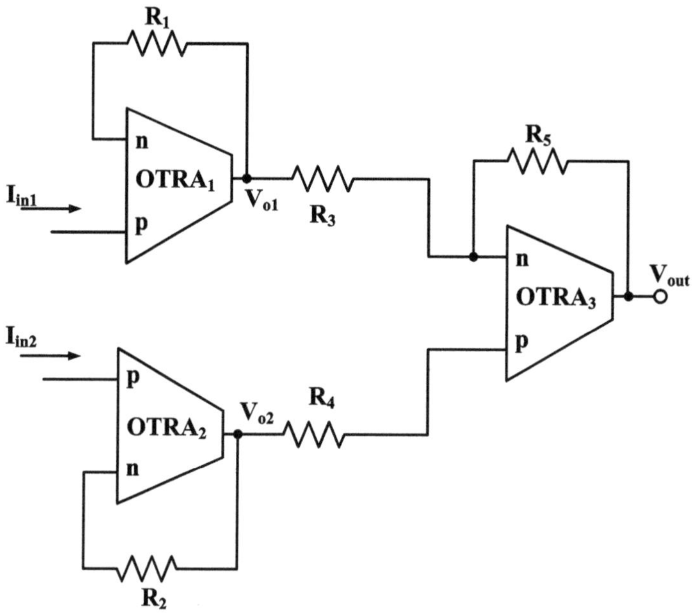

Fig. 5.11 OTRA based I-V CMIA reported in [6]

$$V_{o2} = I_{in2}\left(R_2 + r_{n2}\right) \tag{5.26}$$

By assuming that p and n terminals of $OTRA_3$ are at ground, a simple analysis gives the output voltage as:

$$V_{out} = \left[I_{in2} \frac{R_2 + r_{n2}}{R_4 + r_{p3}} - I_{in1} \frac{R_1 + r_{n1}}{R_3} \right] R_5 \tag{5.27}$$

For differential-mode currents of $I_{in1} = -I_{in2} = I_{dm}/2$, from Eq. (5.27), the differential-mode gain is:

$$A_{dm} = \frac{V_{out}}{I_{dm}} \approx \frac{1}{2}\left[\frac{R_2 + r_{n2}}{R_4 + r_{p3}} + \frac{R_1 + r_{n1}}{R_3} \right] R_5 \tag{5.28}$$

In Eqs. (5.27) and (5.28), r_{n3} is neglected because the negative feedback established by R_5 significantly reduces its value. For common-mode currents of $I_{in1} = I_{in2} = I_{cm}$, from Eq. (5.27), common-mode gain is:

$$A_{cm} = \frac{V_{out}}{I_{cm}} = \left[\frac{\left(R_2 + r_{n2}\right)}{R_4 + r_{p3}} - \frac{\left(R_1 + r_{n1}\right)}{R_3} \right] R_5 \tag{5.29}$$

From Eqs. (5.28) and (5.29), CMRR is found as:

$$CMRR = \frac{\left[R_3\left(R_2 + r_{n2}\right) + \left(R_4 + r_{p3}\right)\left(R_1 + r_{n1}\right)\right]}{2\left[R_3\left(R_2 + r_{n2}\right) - \left(R_4 + r_{p3}\right)\left(R_1 + r_{n1}\right)\right]} \tag{5.30}$$

As Eq. (5.30) indicates, CMIA of Fig. 5.10 requires the matching between the parasitic parameters of r_{n1} and r_{n2} and also matching between the used resistors. However, even in case of full matching between $OTRA_1$ and $OTRA_2$, and the used resistors, since the two inputs of $OTRA_3$ is not symmetrical, the finite value of r_{p3} is the main limiting factor of CMRR. This is the main weakness of this topology.

5.3 Voltage Input-Current Output (V-I) CMIAs

5.3.1 OFCC-Based CMIA

In [4], a V-I CMIA based on OFCC is presented (see Fig. 5.12). The proposed topology comprises of three OFCC and four resistors. $OFCC_1$ and $OFCC_2$ operate as voltage to current convertors. The Z^+ output of $OFCC_1$ and Z^- output of $OFCC_2$ are connected to make a current subtraction node. $OFCC_3$ performs as a current amplifier

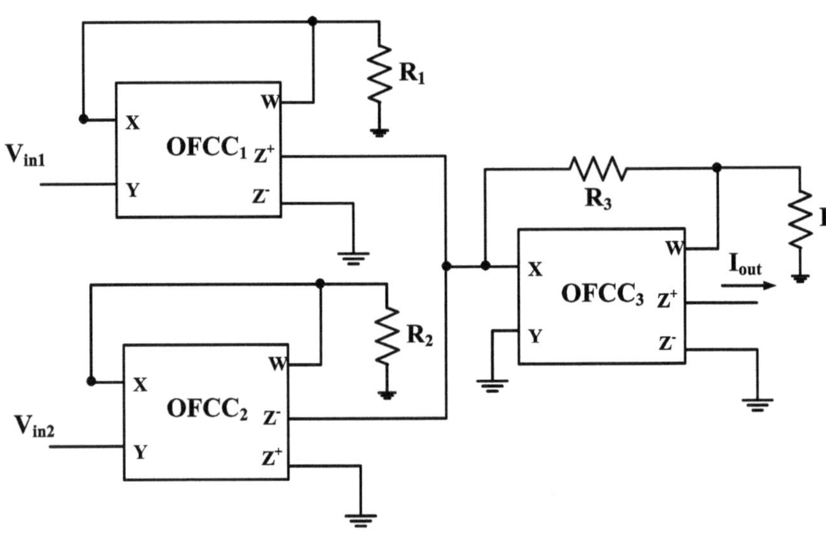

Fig. 5.12 OFCC based V-I CMIA presented in [4]

and amplifies the output current of first stage. Using the operation principles of OFCC in Sect. 5.1.3, analysis of the circuit of Fig. 5.12 gives the output voltage as:

$$I_{out} = \left(\beta_1 \frac{\alpha_1 V_{in1}}{R_1} - \gamma_2 \frac{\alpha_2 V_{in2}}{R_2}\right)\left(1 + \frac{R_3}{R_4}\right)\beta_3 \tag{5.31}$$

In Eq. (5.31), for common-mode inputs of $V_{in1} = V_{in2} = V_{cm}$, common-mode gain is achieved as:

$$A_{cm} = \frac{I_{out}}{V_{in}} = \left(\beta_1 \frac{\alpha_1}{R_1} - \gamma_2 \frac{\alpha_2}{R_2}\right)\left(1 + \frac{R_3}{R_4}\right)\beta_3 \tag{5.32}$$

For differential-mode inputs of $V_{in1} = - V_{in2} = V_{dm}/2$, from Eq. (5.31), differential-mode gain is found as:

$$A_{dm} = \frac{I_{out}}{V_{dm}} = \frac{1}{2}\left(\beta_1 \frac{\alpha_1}{R_1} + \gamma_2 \frac{\alpha_2}{R_2}\right)\left(1 + \frac{R_3}{R_4}\right)\beta_3 \tag{5.33}$$

From Eqs. (5.32) and (5.33), CMRR is calculated as:

$$CMRR = \frac{\beta_1 \frac{\alpha_1}{R_1} + \gamma_2 \frac{\alpha_2}{R_2}}{2\left[\beta_1 \frac{\alpha_1}{R_1} - \gamma_2 \frac{\alpha_2}{R_2}\right]} \tag{5.34}$$

From Eq. (5.34), to have high CMRR, the following condition must be satisfied:

$$\beta_1 \frac{\alpha_1}{R_1} \approx \gamma_2 \frac{\alpha_2}{R_2} \tag{5.35}$$

5.4 Voltage Input-Voltage Output (V-V) CMIAs

5.4.1 OFCC Based V-V CMIAs

Figure 5.13 shows a V-V CMIA topology made of three OFCC and three resistors reported in [4]. In this circuit, $OFCC_1$ and $OFCC_2$ are configured as voltage to current converters. Based on the operation principle of OFCC given in Sect. 5.1.3, their outputs can be expressed as:

$$I_{Z1+} = \alpha_1 \beta_1 \frac{V_{in1}}{R_1},\ I_{Z2-} = \alpha_2 \gamma_2 \frac{V_{in2}}{R_2} \tag{5.36}$$

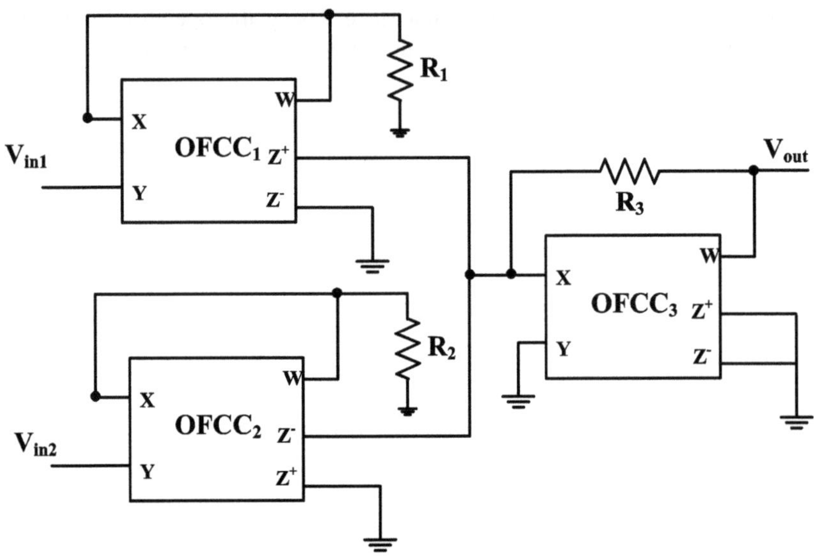

Fig. 5.13 OFCC based V-V CMIA of [4]

$OFCC_3$ is configured as current to voltage converter with gain of R_3. The input current of $OFCC_3$ is $I_{z1+} - I_{z2-}$. Therefore, the output voltage is:

$$V_{out} = \left(\alpha_1 \beta_1 \frac{V_{in1}}{R_1} - \alpha_2 \gamma_2 \frac{V_{in2}}{R_2} \right) R_3 \tag{5.37}$$

For common-mode inputs of $V_{in1} = V_{in2} = V_{cm}$, from Eq. (5.37), common-mode gain is achieved as:

$$A_{cm} = \frac{V_{outc}}{V_{cm}} = \left(\frac{\alpha_1 \beta_1}{R_2} - \frac{\alpha_2 \gamma_2}{R_1} \right) R_3 \tag{5.38}$$

For differential-mode inputs of $V_{in1} = -V_{in2} = V_{dm}/2$, from Eq. (5.37), differential-mode gain is achieved as:

$$A_{dm} = \frac{V_{outd}}{V_{dm}} = \frac{1}{2} \left(\frac{\alpha_1 \beta_1}{R_2} + \frac{\alpha_2 \gamma_2}{R_1} \right) R_3 \tag{5.39}$$

From Eqs. (5.37) and (5.38), CMRR is achieved as:

$$CMRR = \frac{\frac{\alpha_1 \beta_1}{R_2} + \frac{\alpha_2 \gamma_2}{R_1}}{2 \left[\frac{\alpha_1 \beta_1}{R_2} - \frac{\alpha_2 \gamma_2}{R_1} \right]} \tag{5.40}$$

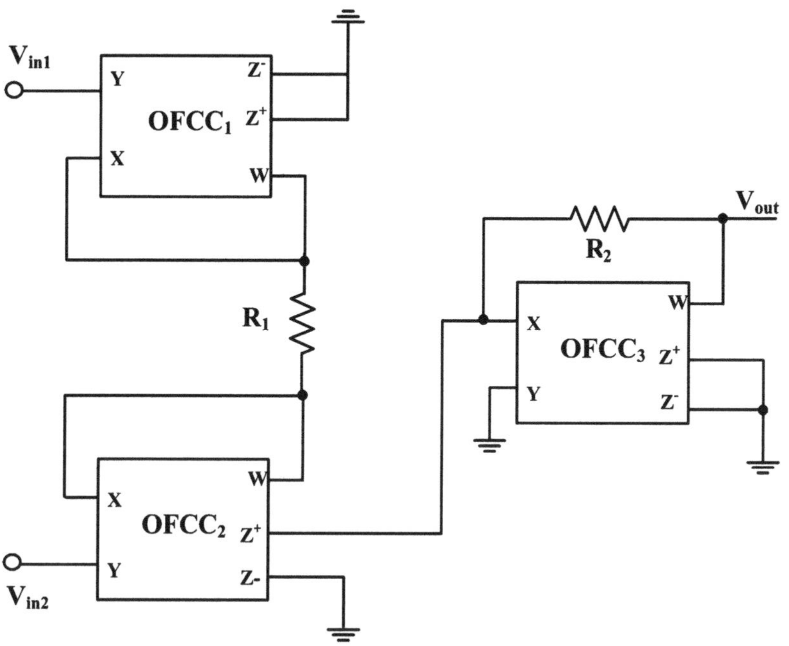

Fig. 5.14 OFCC based V-V CMIA reported in [4]

The CMIA of Fig. 5.13 exhibits high input impedance and low output impedance which is ideal for voltage-input voltage-output applications. However, as Eq. (5.40) indicates, it requires good matching between R_1 and R_2 and also good matching between non-ideal parameters of $OFCC_1$ and $OFCC_2$ to maintain high CMRR.

Other simple OFCC based V-V CMIA topologies are reported in [4, 7, 8]. The CMIA of [4], shown in Fig. 5.14, is capable of achieving higher CMRR compared to the CMIA of Fig. 5.13, because it alleviates the need for any resistor matching. It has also much simpler structure with reduced number of resistors. In this circuit $OFCC_1$ and $OFCC_2$ are configured as voltage buffers and, as a result, the input voltages are copied across R_1. A proportional current is produced in R_1 and is converted to voltage by $OFCC_3$ which is configured as trans-impedance amplifier.

A simple analysis shows the relation between inputs and output voltage as:

$$V_{out} = \frac{\alpha_1 V_{in1} - \alpha_2 V_{in2}}{R_1} \beta_2 R_2 \tag{5.41}$$

For common-mode inputs of $V_{in1} = V_{in2} = V_{cm}$, from Eq. (5.41), common-mode gain is achieved as:

$$A_{cm} = \frac{V_{outc}}{V_{cm}} = \frac{\alpha_1 - \alpha_2}{R_1} \beta_2 R_2 \tag{5.42}$$

For differential-mode inputs of $V_{in1} = -V_{in2} = V_{dm}/2$, from Eq. (5.41), differential-mode gain is achieved as:

$$A_{dm} = \frac{V_{outd}}{V_{dm}} = \frac{\alpha_1 + \alpha_2}{2R_1}\beta_2 R_2 \tag{5.43}$$

From Eqs. (5.42) and (5.43), CMRR is expressed as:

$$CMRR = \frac{\alpha_1 + \alpha_2}{2[\alpha_1 - \alpha_2]} \tag{5.44}$$

As it is seen from Eq. (5.44), matching between only two parameters of α_1 and α_2 results in high CMRR.

5.4.2 *DDCC-Based V-V CMIAs*

Having the property of producing the difference between input voltages makes the differential difference current conveyor (DDCC) a suitable element to implement voltage input CMIAs. The symbol of DDCC is depicted in Fig. 5.15. It has three high impedance voltage input ports (Y_1, Y_2 and Y_3), a low impedance current input port (X) and a high impedance current output port (Z). The relation between terminals voltages and currents are [9]:

$$V_X = \alpha\left(V_{Y1} - V_{Y2} + V_{Y3}\right), \quad I_Z = \beta I_X \tag{5.45}$$

DDCC is employed in [10] to build a V-V CMIA, depicted in Fig. 5.16. The circuit consists of one DDCC, two resistors and one voltage buffer. Input signals are applied to Y_1 and Y_2 nodes. The difference between V_{Y1} and V_{Y2} are copied to X node where it is converted to current by R_1. The current produced at R_1 is transferred to Z node and converted to voltage by R_2. The output voltage is expressed as:

$$V_{out} = \alpha\beta\left(V_{in1} - V_{in2}\right)\frac{R_2 \parallel r_z}{R_1 + r_x} \tag{5.46}$$

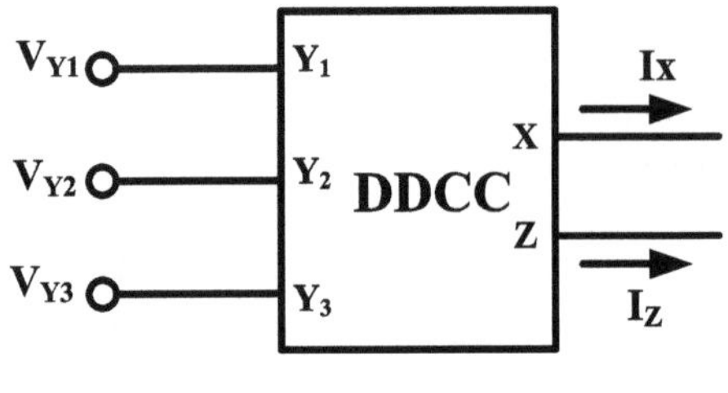

Fig. 5.15 Schematic representation of DDCC [9]

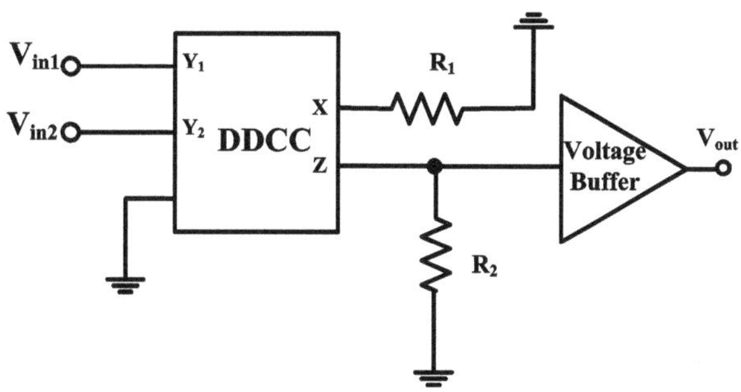

Fig. 5.16 DDCC based V-V CMIA reported in [9]

where r_x and r_z are parasitic impedances at X and Z terminals respectively. In the DDCC based CMIA topology of Fig. 5.16 only one active element is used and no matching between resistors is required. The final value of CMRR is determined by the performance of DDCC and how well it can produce the difference between input voltages. Generally, the voltage differencing stage is implemented by a differential pair and its CMRR is very important parameter which determines the overall CMRR value of DDCC-based CMIA.

A method is reported in [9] to improve CMRR of Fig. 5.16 by processing common-mode input signals using another DDCC component as is shown in Fig. 5.17. The common-mode component of input signals (V_{cm}) is extracted by two resistors R_{cm1} and R_{cm2} and applied to the Y_1 and Y_2 inputs of $DDCC_2$. Input voltage buffers are used to isolate input signals from R_{cm1} and R_{cm2}. Since equal voltages are fed into the Y_1 and Y_2 terminals of $DDCC_2$, according to Eq. (5.46), the output signal should be zero if Y_3 voltage is zero. If there is a non-zero output level at Z terminal, negative feedback loop, established by $DDCC_2$, feeds back an appropriate voltage to the Y_3 terminal of $DDCC_2$ to keep the output at Z terminal of $DDCC_2$ equal to zero. The same voltage is also fed to Y_3 terminal of $DDCC_1$ to cancel the effect of common-mode input voltages at Z terminal. Common-mode feedback in the CMIA of Fig. 5.17 provides a very high CMRR compared to the CMIA of Fig. 5.16. Frequency compensation may be required in CMIA of Fig. 5.17. Another DDCC based implementation of V-V CMIA is reported in [10] and shown in Fig. 5.18 in which $DDCC_1$ and $DDCC_2$ are configured to boost the differential-mode gain and CMRR. Any offset is also cancelled by $DDCC_3$ and $DDCC_4$. In this circuit, the final value of CMRR is limited by mismatch between $DDCC_1$ and $DDCC_2$, as well as mismatch between the two resistors named R_2.

Figure 5.19a shows another DDCC based V-V CMIA implementation reported in [11]. For practical use, the circuit requires an extra voltage buffer at the output. By assuming unity voltage and current transfer gains for DDCC, the output voltage is:

$$V_{out} \approx \left(V_{in1} - V_{in2}\right)\frac{R_2 \parallel r_z}{R_1 + r_x} \tag{5.47}$$

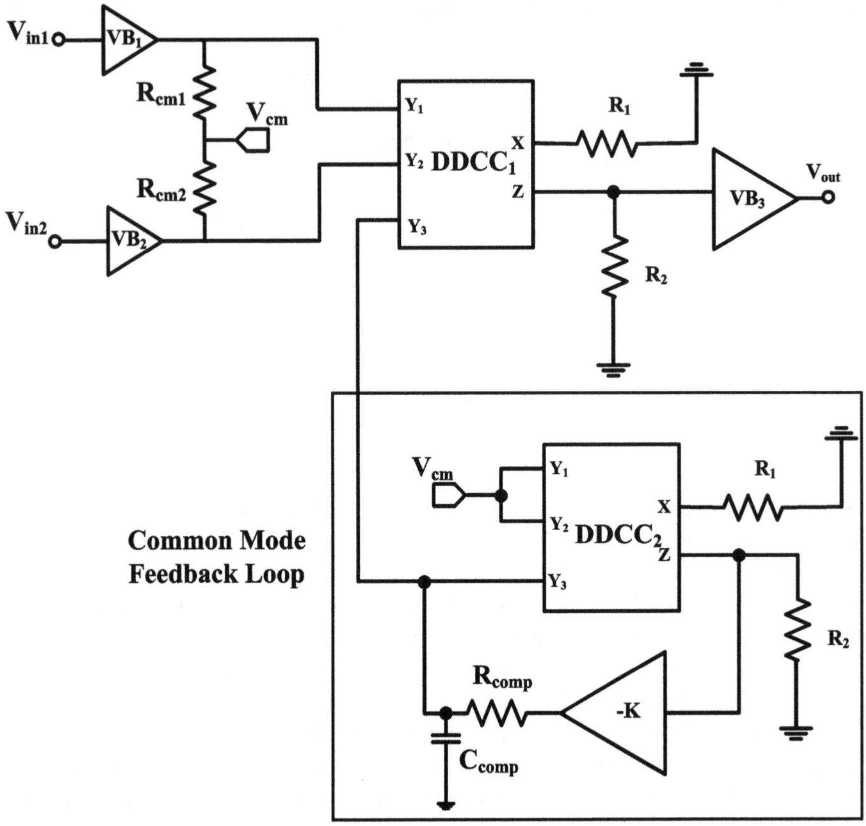

Fig. 5.17 High CMRR DDCC based V-V CMIA with common-mode feedback reported in [9]

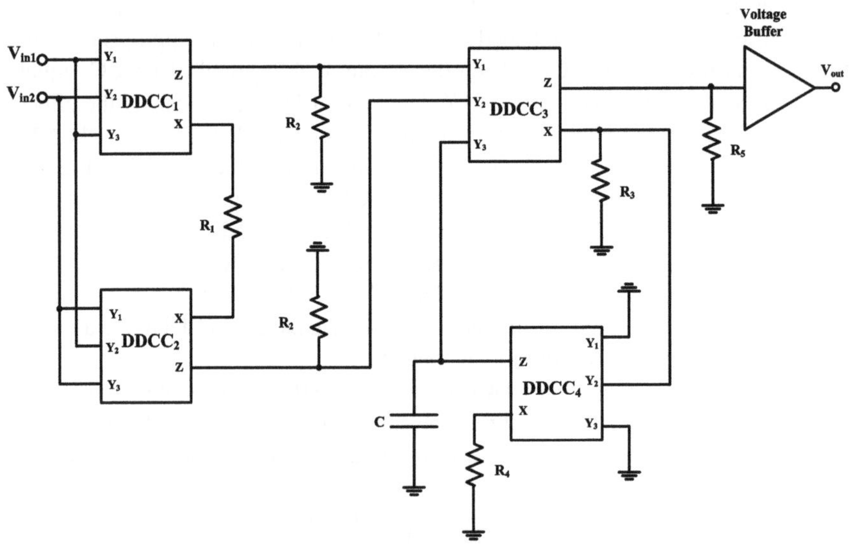

Fig. 5.18 DDCC based V-V CMIA with offset cancellation reported in [10]

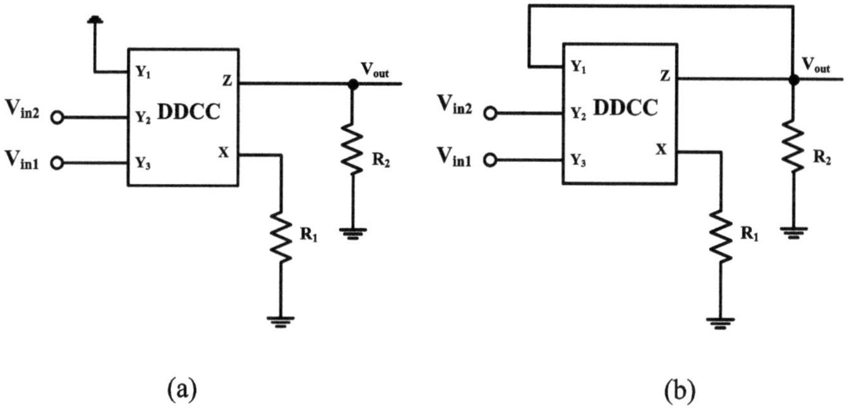

Fig. 5.19 DDCC based V-V CMIA (**a**) simple implementation (**b**) with improved gain error [11]

According to Eq. (5.48), to reduce the gain error caused by r_x variations, we must have $R_1 \gg r_x$. However, to increase differential gain, we need to reduce the value of R_1. Gain can also be increased by increasing the value of R_2 which is limited by the parasitic impedance at Z terminal. Therefore, there is a trade-off between gain error and the ultimate value of achievable gain. Figure 5.19b shows a modified realization of DDCC based V-V CMIA of Fig. 5.19a which is capable of achieving higher gain [11]. By assuming unitary voltage and current gains for DDCC, the output voltage is expressed as:

$$V_{out} = \left(V_{in1} - V_{in2}\right)\frac{R_2 \parallel r_z}{R_1 + r_x - R_2 \parallel r_z} \tag{5.48}$$

5.4.3 DVCC-Based V-V CMIAs

Another member of current conveyors family is the Differential Voltage Current Conveyor (DVCC), shown in Fig. 5.20. It is a four terminal device able to convey the difference between input voltages at Y_1 and Y_2 terminals to X terminal. The operation is similar to DDCC device with Y_3 terminal grounded. The relations between terminals voltage and currents are [12]:

$$V_x = \alpha_1 V_{in1} - \alpha_2 V_{in2}, \quad I_Z = \pm\beta I_x \tag{5.49}$$

The + and – signs are representing DVCC$^+$ and DVCC$^-$, respectively. Figure 5.21 shows a simple implementation of DVCC based V-V CMIA reported in [12] which utilizes the voltage differencing property of DVCC. Its operation is similar to the CMIA of Fig. 5.16 and requires a voltage buffer at output. A simple analysis gives

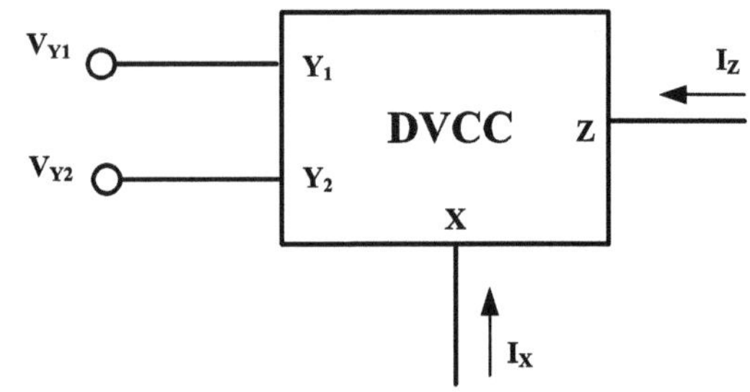

Fig. 5.20 Schematic representation of DVCC [12]

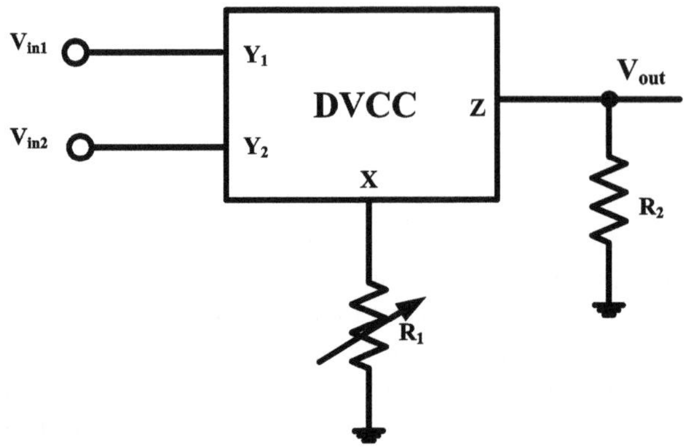

Fig. 5.21 DVCC based V-V CMIA reported in [12]

the differential-mode gain (for $V_{in1} = -V_{in2} = V_{dm}/2$), common-mode gain (for $V_{in1} = V_{in2} = V_{cm}$) and CMRR as respectively:

$$A_{dm} = \frac{V_{outd}}{V_{dm}} = \frac{1}{2}(\alpha_1 + \alpha_2)\beta\frac{R_2}{R_1},\quad A_{cm} = \frac{V_{outc}}{V_{cm}} = (\alpha_1 - \alpha_2)\beta\frac{R_2}{R_1} \tag{5.50}$$

$$CMRR = \frac{A_{dm}}{A_{cm}} = \frac{1}{2}\frac{\alpha_1 + \alpha_2}{\alpha_1 - \alpha_2} \tag{5.51}$$

According to Eq. (5.51), the matching between α_1 and α_2 determines the final value of CMRR.

Figure 5.22 shows another V-V CMIA structure reported in [13] which is made of two DVCC blocks. The circuit is capable of achieving higher differential-mode gain and CMRR compared to CMIA of Fig. 5.21. The relation between input voltages and output voltage is [13]:

$$V_{out} = \beta_1(1+\beta_2)\left[(\alpha_2 + \alpha_1')V_{in2} - (\alpha_2' + \alpha_1)V_{in1}\right]\frac{R_2}{R_1} \tag{5.52}$$

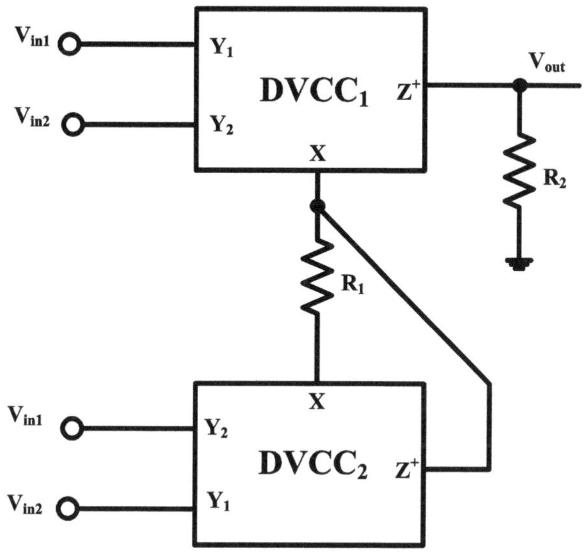

Fig. 5.22 V-V CMIA based on two DVCC blocks reported in [13]

where β_i and α_i (for i = 1,2) are non-ideal parameters of $DVCC_1$ and β'_i and α'_i (for i = 1,2) are non-ideal parameters of $DVCC_2$.

For $V_{in1} = -V_{in2} = V_{dm}/2$, using Eq. (5.52) the differential-mode gain is:

$$A_{dm} = \beta_1\left(1+\beta_2\right)\left[\left(\alpha_2+\alpha_1'\right)+\left(\alpha_2'+\alpha_1\right)\right]\frac{R_2}{2R_1} \tag{5.53}$$

For $V_{in1} = V_{in2} = V_{cm}$, using Eq. (5.52) the common-mode gain is:

$$A_{cm} = \beta_1\left(1+\beta_2\right)\left[\left(\alpha_2+\alpha_1'\right)-\left(\alpha_2'+\alpha_1\right)\right]\frac{R_2}{R_1} \tag{5.54}$$

From Eqs. (5.53) and (5.54), CMRR is found as [13]:

$$CMRR = \frac{\left(\alpha_2+\alpha_1'\right)+\left(\alpha_2'+\alpha_1\right)}{2\left[\left(\alpha_2+\alpha_1'\right)-\left(\alpha_2'+\alpha_1\right)\right]} = \frac{\left(\alpha'_2+\alpha_1'\right)+\left(\alpha_2+\alpha_1\right)}{2\left[\left(\alpha_1'-\alpha_2'\right)-\left(\alpha_1-\alpha_2\right)\right]} \tag{5.55}$$

Compared to the single DVCC based CMIA of Fig. 5.21, the CMIA of Fig. 5.22 exhibits higher differential-mode gain. In addition, as the denominator of Eq. (5.55) indicates, if the non-ideal voltage gains of $DVCC_1$ and $DVCC_2$ show similar variation trend, a high value of CMRR can be achieved [13].

5.4.4 CFOA-Based V-V CMIAs

Figure 5.23 shows a CMIA topology based on only one CFOA and two resistors reported in [3]. Although it uses only one active building block, it suffers from unequal input impedances. The input impedance seen by V_{in1} is high but is equal to R_1 for V_{in2} input. The output impedance is low and no extra voltage buffer is required. Based on the CFOA characteristics discussed in Sect. 5.1.2, a routine analysis gives the output voltage as:

$$V_{out} = \left(\alpha_1 V_{in1} - V_{in2}\right)\beta\alpha_2 \frac{R_2}{R_1} \tag{5.56}$$

For differential-mode inputs ($V_{in1} = -V_{in2} = V_{dm}/2$), from Eq. (5.56), differential-mode gain is found as:

$$A_{dm} = \frac{V_{out}}{V_{dm}} = \frac{1}{2}\left(\alpha_1 + 1\right)\beta\alpha_2 \frac{R_2}{R_1} \tag{5.57}$$

For common-mode inputs ($V_{in1} = V_{in2} = V_{cm}$), common-mode gain is expressed as:

$$A_{cm} = \frac{V_{out}}{V_{cm}} = \left(\alpha_1 - 1\right)\beta\alpha_2 \frac{R_2}{R_1} \tag{5.58}$$

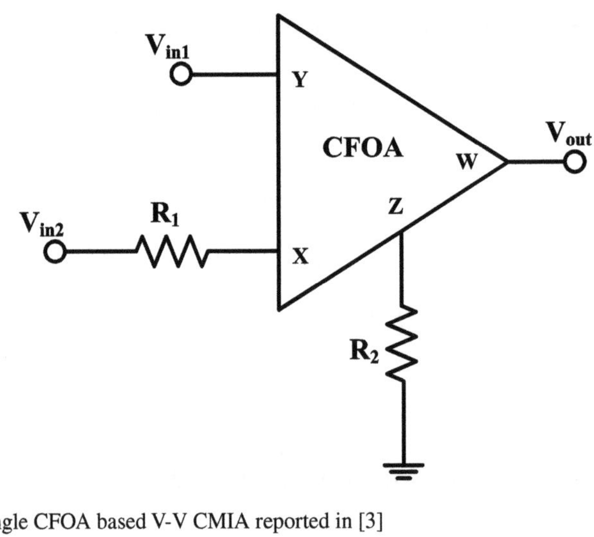

Fig. 5.23 Single CFOA based V-V CMIA reported in [3]

From Eqs. (5.57) and (5.58), CMRR is expressed as:

$$CMRR = \frac{A_{dm}}{A_{cm}} = \frac{\alpha_1 + 1}{2(\alpha_1 - 1)} \tag{5.59}$$

Equation (5.59) states that the final value of CMRR is determined by α_1 value. Figure 5.24 shows another CFOA based V-V CMIA reported in [3] where the value of CMRR can be set by ratio of resistors R_2/R_1. It employs two CFOAs and four resistors and unlike the CMIA of Fig. 5.23, it has equal impedances for both inputs. In this circuit the CMRR is found to be:

$$CMRR = \frac{A_{dm}}{A_{cm}} = \frac{\left(\alpha'_1 + \beta_1\alpha_1\alpha_2\frac{R_2}{R_1}\right)}{2\left(\alpha'_1 - \beta_1\alpha_1\alpha_2\frac{R_2}{R_1}\right)} \tag{5.60}$$

where α_1, α_2 and β_1 parameters are related to $CFOA_1$ and α'_1 belongs to $CFOA_2$.

In Eq. (5.60), a very high value is achievable for CMRR if the following condition is met:

$$\alpha'_1 = \beta_1\alpha_1\alpha_2\frac{R_2}{R_1} \tag{5.61}$$

Therefore, by adjusting the ratio of R_2/R_1, the denominator of Eq. (5.61) can be made very low resulting in a very high CMRR.

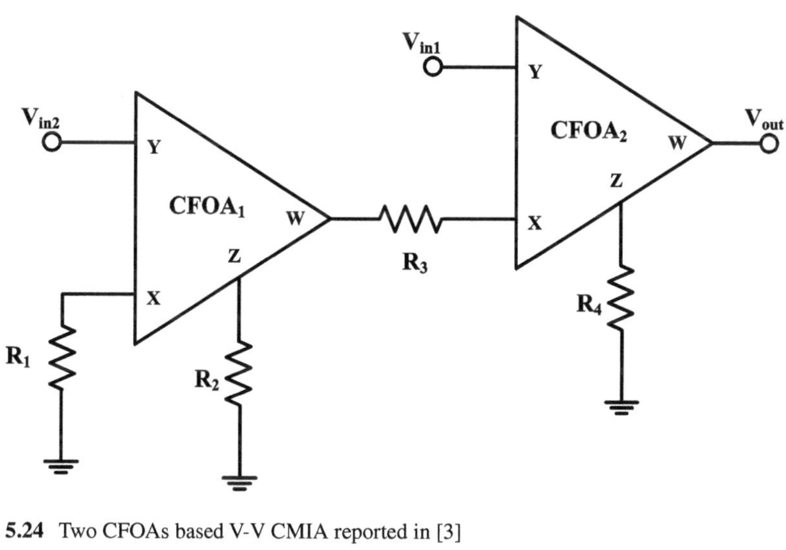

Fig. 5.24 Two CFOAs based V-V CMIA reported in [3]

References

1. Gupta K., Gupta P., Pandey N., Pandey R. (2016) CDBA – current based instrumentation amplifier. Journal of Communications Technology, Electronics and Computer Science, 4:11–15.
2. Acar C., Ozuguz S. (1999) A new versatile building block: current differencing buffered amplifier suitable for analog processing filters. Microelectronics Journal, 30:157–160.
3. Yuce E. (2010) Various current-mode and voltage-mode instrumentation amplifier topologies suitable for integration. Journal of Circuits, Systems, and Computers 19(3):89–699.
4. Pandey N., Nand D., Pandey R. (2016) Generalized operational floating current conveyor based instrumentation amplifier.IET Circuits Devices Systems, 10:209–219.
5. Pandey N., Nand D., Kumar V. V., Ahalawat V. K., Malhotra C. (2016) Realization of OFCC based Transimpedance mode instrumentation amplifier.Theoretical and Applied Electrical Engineering, 14(2):162–167.
6. Pandey R., Pandey N., Paul K. (2013) Electronically tunable transimpedance instrumentation amplifier based on OTRA. Journal of Engineering, 10:1–5, 2013.
7. Nand D., Pandey N. (2017) A new proposal for OFCC-based instrumentation amplifier. International Journal of Electrical and Computer Engineering (IJECE), 7(1):134–143.
8. Ghallab Y. H., Badawy W, Kaler K. V. I. S., Maundy B. J. (2005) A novel current-mode instrumentation amplifier based on operational floating current conveyor. IEEE Transactions on Instrumentation and Measurement, 54(5):1941–1949.
9. Cini U. (2014) A low-offset high CMRR current-mode instrumentation amplifier using differential difference current conveyor.IEEE International Conference on Electronics, Circuits and Systems (ICECS), 2014.
10. Cini U., Arslan E. (2015) A high gain and low-offset current-mode instrumentation amplifier using differential difference current conveyors. IEEE International Conference on Electronics, Circuits, and Systems (ICECS), 2015.
11. Yuce E. (2014) Novel instrumentation amplifier and integrator circuits using single DDCC and only grounded passive elements. Indian journal of Pure & Applied Physics, 52:767–757.
12. Hassan T., Mahmoud S. A. (2010) New CMOS DVCC realization and applications to instrumentation amplifier and active-RC filters. International Journal of Electronics and Communications (AEÜ), 64:47–55.
13. Yang T. Y., Liu K. Y., Wang H. Y. (2011) Novel high-CMRR DVCC-based instrumentation amplifier. International Conference on Engineering and Industries (ICEI), 2011.

Chapter 6
Electronically Controllable Current-Mode Instrumentation Amplifiers

6.1 CMIA Based on Electronically Controllable Current-Mode Active Building Blocks

6.1.1 Current Controlled Differential Voltage Current Conveyor Based CMIA

The Current Controlled Differential Voltage Current Conveyor (CCDVCC) is a member of current conveyor family where parasitic resistance at X terminal can be changed electronically. This feature makes it possible to use the intrinsic *X* terminal parasitic resistor in variable gain applications instead of using external passive resistors. Figure 6.1 shows symbol of CCDVCC in which parasitic resistor at *X* terminal is varied by I_B current source. Y_1 and Y_2 are high impedance voltage input terminals, *X* is current input terminal and Z_1, Z_2 are high impedance current output terminals. The relations between terminals voltages and currents are the following [1]:

$$V_x = \alpha\left(V_{Y1} - V_{Y2}\right), \quad V_x = I_x r_x, \quad r_x = f\left(I_B\right), \quad I_{z1} = -I_{Z2} = \beta I_x \tag{6.1}$$

where r_x, α and β are: parasitic resistance at X terminal (which is a function of I_B), voltage gain in transferring the difference $V_{Y1}-V_{Y2}$ to X terminal and current gain between X and Z terminals, respectively. The values of α and β are typically unitary.

Figure 6.2 shows the CCDVCC-based electronically controllable CMIA reported in [1]. It is simply formed by connecting resistors at Z terminals; however, for practical application it needs voltage buffers at the output nodes. In this circuit, the difference between input voltages ($V_{Y1}-V_{Y2}$) is conveyed to X terminal where it is converted to a proportional current by its parasitic resistance r_x:

G. Ferri et al., *Current-Mode Instrumentation Amplifiers*, Analog Circuits and Signal Processing, https://doi.org/10.1007/978-3-030-01343-1_6

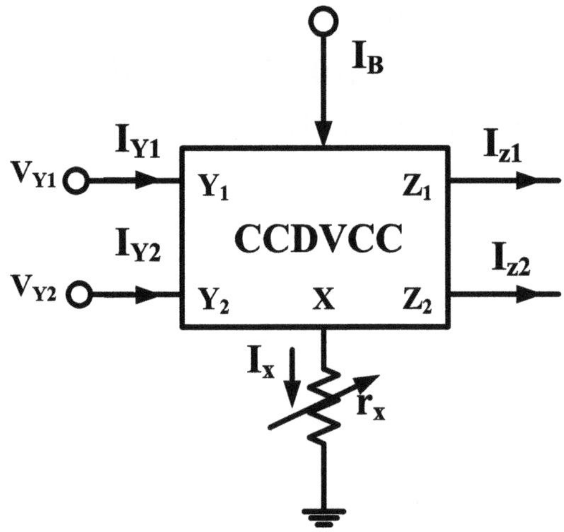

Fig. 6.1 Schematic representation of CCDVCC [1]

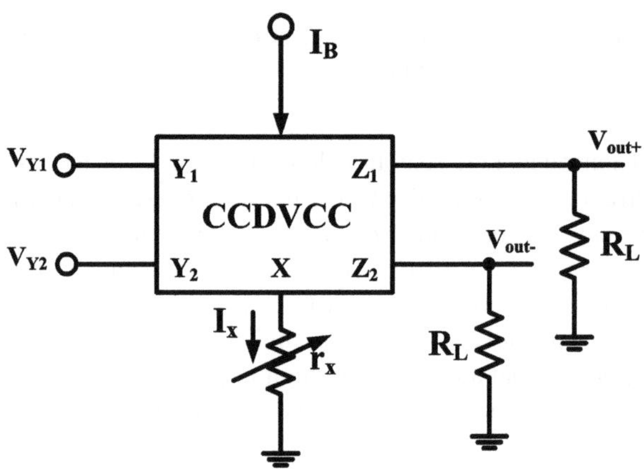

Fig. 6.2 CCDVCC based CMIA of [1]

$$I_x = \alpha \frac{V_{Y1} - V_{Y2}}{r_x} \tag{6.2}$$

The current at X terminal is transferred to Z terminals where it is converted to voltage by R_L as:

$$V_{out+} = -V_{out-} = \alpha\beta \frac{R_L}{r_x}\left(V_{Y1} - V_{Y2}\right) \tag{6.3}$$

A possible CMOS implementation of CCDVCC reported in [1] is shown in Fig. 6.3. In this circuit, voltage difference is produced by M_1-M_6, while electronically controllable resistor at X terminal is achieved by M_8-M_{11}. Transistors M_{12}-M_{21}

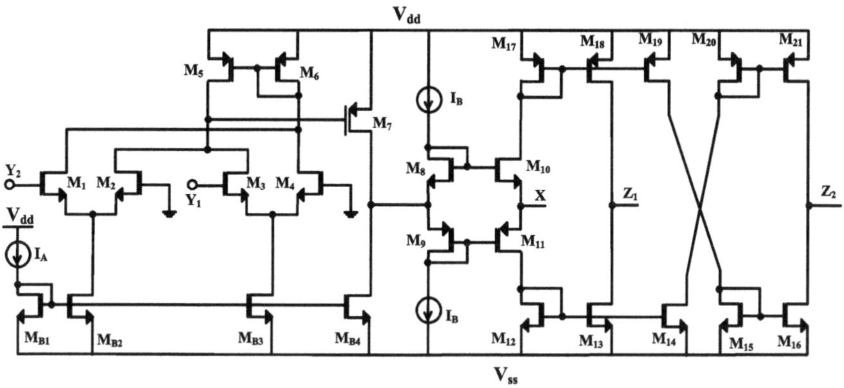

Fig. 6.3 CMOS implementation of CCDVCC [1]

are used for Z_1 and Z_2 outputs. By assuming matching between M_{10} and M_{11}, the impedance at X terminal is expressed as [1]:

$$r_x = \frac{1}{\sqrt{8KI_B}} \tag{6.4}$$

where K is defined as (with usual of meaning of symbols):

$$K = \mu C_0 \frac{W}{L} \tag{6.5}$$

Inserting Eq. (6.4) into Eq. (6.3) and considering the usual definition of A_{dm}, gives the differential-mode gain as:

$$A_{dm} = \alpha\beta R_L \sqrt{8KI_B} \tag{6.6}$$

According to Eq. (6.6), higher gain is obtained by increasing I_B. However, a high value of I_B results in larger power consumption. In addition, increasing I_B reduces output impedances at Z_1 and Z_2 outputs limiting the maximum value of load resistors connected to these terminals. Consequently, the maximum value of achievable gain is limited.

6.1.2 CMIA Based on Current Controlled Current Conveyor Trans-conductance Amplifier

Another member of current conveyor family used in the implementation of electronically tunable CMIAs is the Current Controlled Current Conveyor Transconductance Amplifier (CCCCTA) which is a four-terminal device. Figure 6.4 shows (a) the symbol and (b) the equivalent circuit of CCCCTA [2–4]. The relation between terminals voltage and current signals of CCCCTA is:

$$V_x = r_x I_x + \alpha V_y, \quad I_z = \beta I_x, \; I_o = g_m V_z \tag{6.7}$$

In this device, the value of parasitic resistance at X terminal can be controlled by bias current I_{B1}. This feature alleviates the need for external resistor in practical applications. The value of g_m is also controllable by another bias current I_{B2}, as it will be shown, making the output signal less sensitive to temperature variation.

Figure 6.5 shows a CMIA structure based on CCCCTA [3]. In this circuit only two active building blocks are used. The parameters of $CCCCTA_1$ and $CCCCTA_2$ are controlled by I_{B1}-I_{B2} and I_{B3}-I_{B4}, respectively. The CMIA of Fig. 6.5 can produce output signals in both forms of voltage and current simultaneously.

A straight forward analysis yields I_{Z1} as:

$$I_{Z1} = -\beta_1 \frac{\alpha_1 V_1 - \alpha_2 V_2}{r_{x1} + r_{x2}} \tag{6.8}$$

where α_1 and α_2 are the voltage gain and r_{x1} and r_{x2} are the X terminal parasitic resistances of $CCCCTA_1$ and $CCCCTA_2$, respectively.

By considering $I_{O1} = -I_{Z1}$ and $I_{O1} = g_{m1} V_{out}$, the output voltage is found as:

$$V_{out} = \beta_1 \frac{\alpha_1 V_1 - \alpha_2 V_2}{g_{m1}\left(r_{x1} + r_{x2}\right)} \tag{6.9}$$

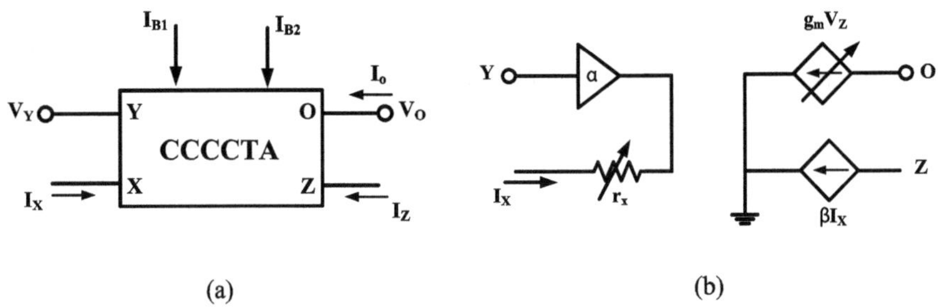

Fig. 6.4 CCCCTA (**a**) symbol (**b**) equivalent circuit [2]

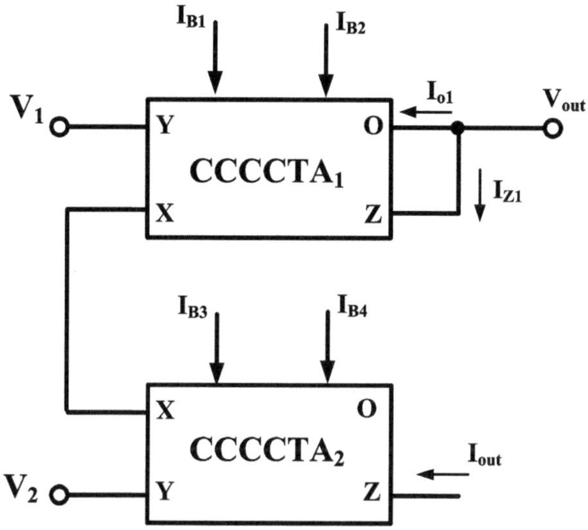

Fig. 6.5 CCCCTA based CMIA of [3]

Using Eq. (6.9), differential-mode gain for differential-mode inputs of $V_1 = -V_2 = V_{dm}/2$ and common-mode gain for common-mode inputs of $V_1 = V_2 = V_{cm}$ are expressed as:

$$A_{dm} = \frac{V_{outd}}{V_{dm}} = \frac{1}{2}\beta_1 \frac{\alpha_1 + \alpha_2}{g_{m1}\left(r_{x1} + r_{x2}\right)} \tag{6.10}$$

$$A_{cm} = \frac{V_{outc}}{V_{cm}} = \beta_1 \frac{\alpha_1 - \alpha_2}{g_{m1}\left(r_{x1} + r_{x2}\right)} \tag{6.11}$$

Using Eqs. (6.10 and 6.11), CMRR is:

$$CMRR = \frac{\alpha_1 + \alpha_2}{2\left(\alpha_1 - \alpha_2\right)} \tag{6.12}$$

According to Eq. (6.12), the ultimate value of CMRR is determined by matching between α_1 and α_2 parameters. In addition, as there are two electronically variable parameters in the denominator of Eq. (6.10), differential-mode gain can be electronically controlled.

Figure 6.6 shows a BICMOS implementation of CCCCTA reported in [3] where, referring to Fig. 6.5, r_X and g_m are:

$$r_{x1} = \frac{V_T}{2I_{B1}}, \quad g_{m1} = \frac{I_{B2}}{2V_T} \tag{6.13}$$

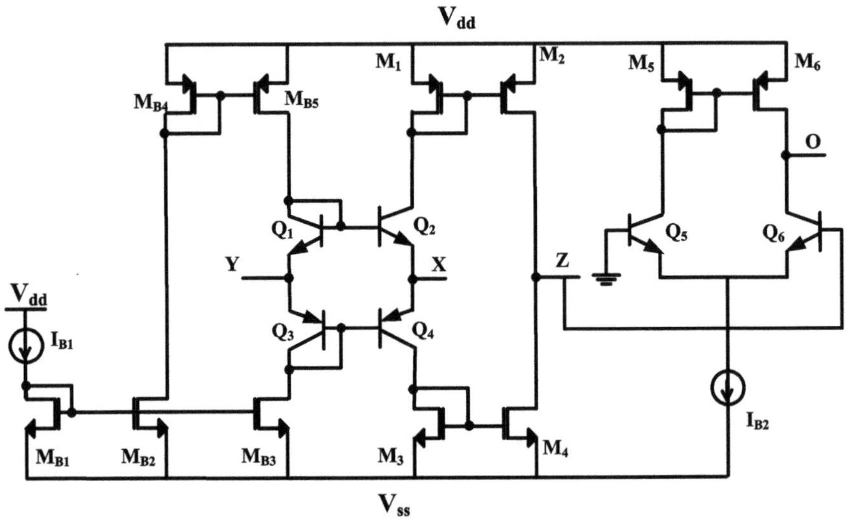

Fig. 6.6 BICMOS implementation of CCCCTA [2]

where V_T is thermal voltage. Inserting Eq. (6.13) into Eq. (6.10), and assuming symmetry between $CCCCTA_1$ and $CCCCTA_2$, gives the differential-mode gain as:

$$A_{dm} = \beta_1 (\alpha_1 + \alpha_2) \frac{I_{B1}}{I_{B2}} \tag{6.14}$$

From Eq. (6.14), if active building blocks parameters (α_1, α_2) are designed to be not dependent on temperature, the gain can be also temperature insensitive because it is independent from V_T. In addition, it can be controlled by current sources.

The CMIA of Fig. 6.5 can also produce an output signal in the form of current which is, on the contrary, temperature sensitive:

$$I_{out} = \beta_2 \frac{\alpha_1 V_1 - \alpha_2 V_2}{r_{x1} + r_{x2}} \tag{6.15}$$

6.1.3 CMIA Based on Current Controlled Current Conveyors

In 1996, Fabre et al. introduced second generation current controlled current conveyor (CCCII) with capability of controlling parasitic resistance at X terminal by means of its bias current [5]. They showed that instead of using passive resistors to change the gain, many applications such as voltage-current convertor, voltage amplifier, current amplifier etc. can take advantage of the resistance at X terminal to

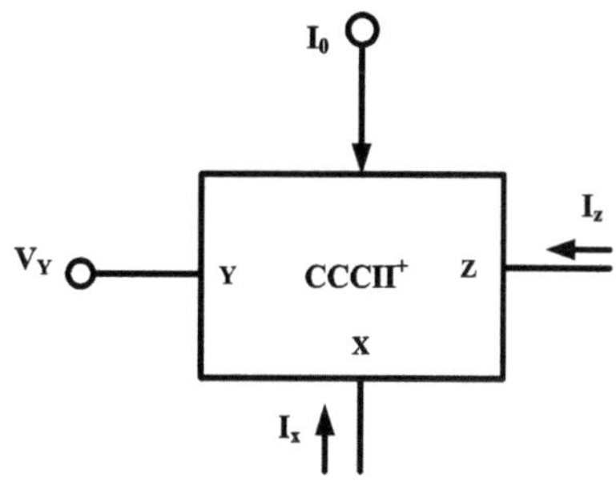

Fig. 6.7 Symbolic representation of CCCII+ [5]

present electronic tunability. Later, this device was employed in [6–8] to realize electronically controllable CMIA.

Figure 6.7 shows the symbolic representation of CCCII+ where parasitic resistance at X terminal, r_x, is controlled by I_0 [5]. In general form, the relations between terminals voltage and current signals are expressed as:

$$V_X = \alpha V_Y + r_x I_x, \quad I_z = \pm \beta I_x \tag{6.16}$$

where α is voltage gain between Y and X terminals and β is current gain between Z and X terminals. For $I_z = +\beta I_x$ we have CCCII+ and $I_z = -\beta I_x$ indicates CCCII−. Figure 6.8 shows simple realizations of CCCII+ and CCCII− reported in [5, 8] using bipolar technology.

Figure 6.9 shows a CCCII based electronically controllable CMIA introduced in [6]. It is based on three CCCIIs and can produce output signals in both the forms of voltage and current at the same time. The Z outputs of $CCCII_1$ and $CCCII_2$ are connected to form a current subtraction node. This connection boosts the differential gain. The output current produced by the first stage is converted to a voltage by parasitic resistance at X terminal of $CCCII_3$. For voltage output, $CCCII_3$ operates as a variable resistor. For current output, $CCCII_3$ provides high output impedance. If voltage output is not required, $CCCII_3$ can be omitted. Otherwise for practical applications, a voltage buffer is required at voltage output. The voltage and current relations between output and input signals are the following:

$$V_{out} = \frac{r_{x3}}{r_{x1} + r_{x2}} (\beta_1 + \beta_2)(\alpha_1 V_1 - \alpha_2 V_2) \tag{6.17}$$

$$I_{out} = \frac{1}{r_{x1} + r_{x2}} (\beta_1 + \beta_2)(\alpha_1 V_1 - \alpha_2 V_2) \tag{6.18}$$

being r_{xi} for i = 1–3, the intrinsic resistance at X node of the related CCCII.

A simplified version of the CMIA of Fig. 6.9 was introduced in [7] where $CCCII_3$ is replaced by an electronically controllable active resistor made of two transistors and all the used active building blocks are of the same type. The CCCII based CMIA of [7], shown in Fig. 6.10, has the advantage of a reduced power consumption and

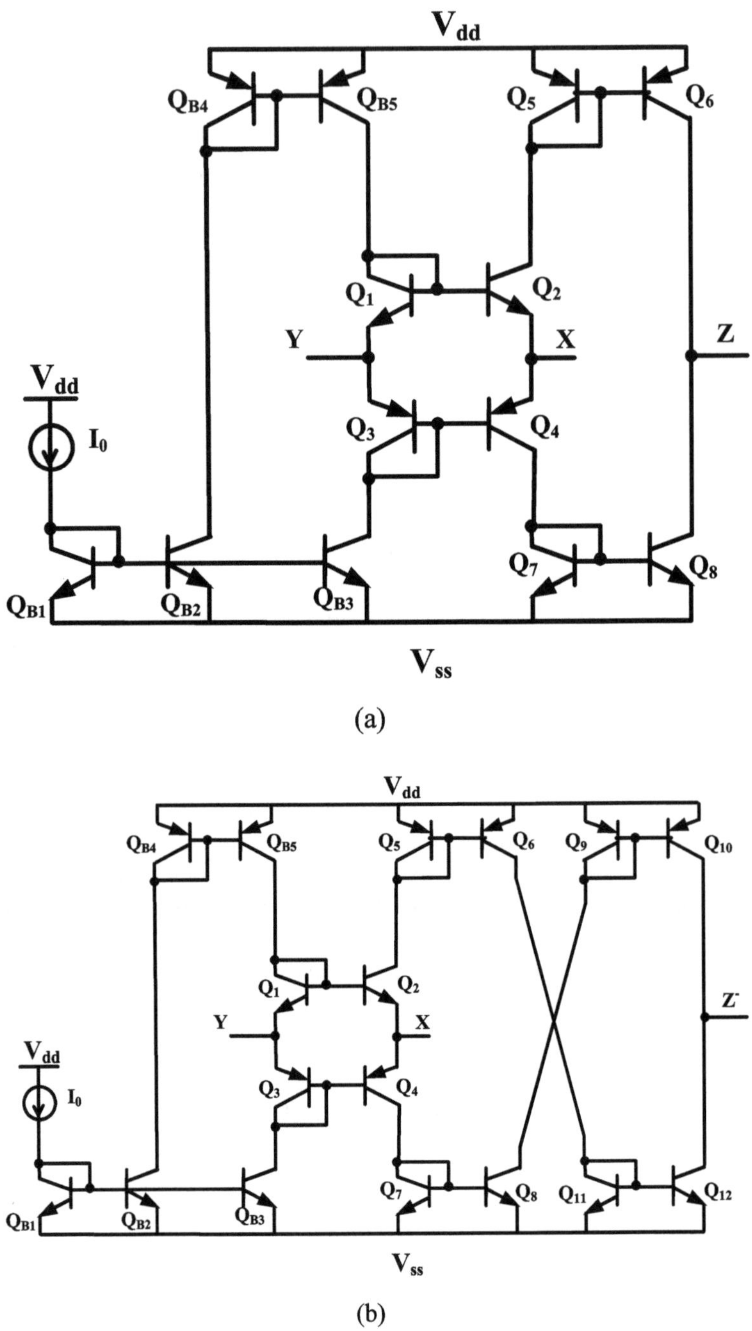

Fig. 6.8 Internal circuit of (**a**) CCCII+ [5] (**b**) CCCII− [9]

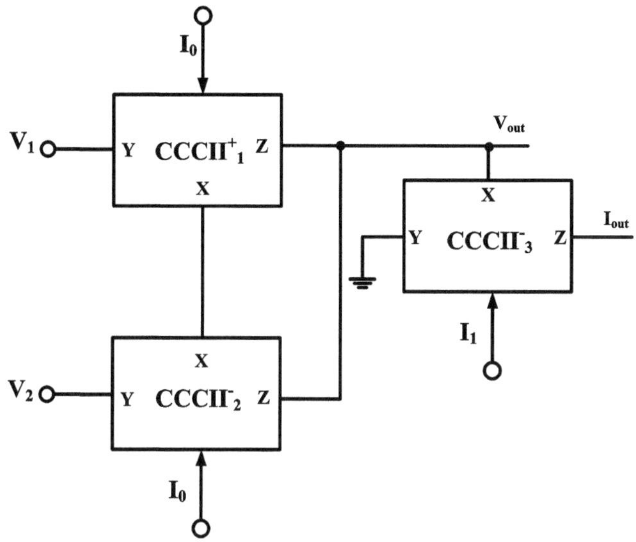

Fig. 6.9 CCCII based CMIA of [6]

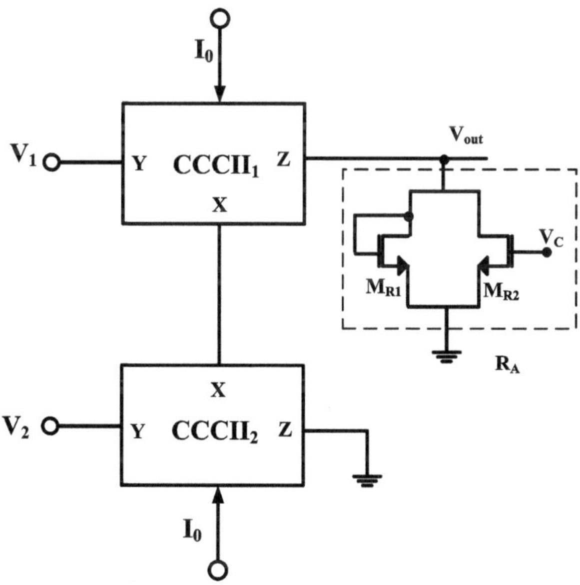

Fig. 6.10 CCCII based CMIA of [7]

chip area if compared to the CMIA of [7]. It employs two CCCIIs and an active resistor which is used to convert output current of $CCCII_1$ into an output voltage. The value of active resistor R_A is a function of control voltage V_C as:

$$R_A = \frac{1}{\mu_n C_{ox}(W/L)(V_C - 2V_{TH})} \tag{6.19}$$

where μ_n, C_{ox} V_{TH} are parameters of M_{R1}-M_{R2} and W/L is their aspect ratios. Equation (6.19) is achieved by assuming that M_{R1} and M_{R2} operate in triode region. However as the gate of M_{R1} is connected to its drain, this relation is valid for depletion mode MOSFETs. The relation between input and output voltages are:

$$V_{out} = \beta_1(\alpha_1 V_1 - \alpha_2 V_2)\frac{R_A}{r_{x1} + r_{x2}} \tag{6.20}$$

According to Eq. (6.19) and Eq. (6.20), gain can be controlled by either V_C or I_0. The final value of CMRR is also determined by the matching between $CCCII_1$ and $CCCII_2$. For practical applications, also this circuit requires a voltage buffer at the output.

Figure 6.11 shows the approach adopted in [8] to implement a CCCII-based CMIA in which instead of an external resistor, the r_x of a third CCCII is used as variable resistor. Compared to the CMIA of Fig. 6.9, it has a simpler design because all the used CCCIIs are of the same type. However, its differential-mode gain is half of the CMIA of Fig. 6.9. In comparison to the CMIA of Fig. 6.10, the simpler implementation of the CMIA of Fig. 6.11 is achieved at the expense of a reduced differential-mode gain and a high power consumption.

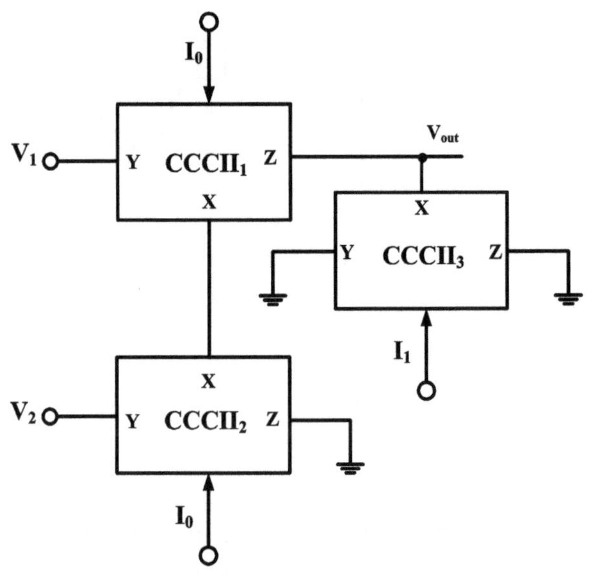

Fig. 6.11 CCCII based CMIA of [8]

6.1.4 CMIA Based on Variable Gain Current Mirror

Another possibility to have an electronically tunable gain is the use of a variable gain current mirror. Figure 6.12 shows the complete schematic of an electronically controllable CMIA based on this method proposed by Safari and Minaei in 2013 [10]. The circuit is formed by three stages. The input stage is a differential current mirror (M_1-M_5) that produces an output current (i_{o1}) proportional to the difference between the input currents i_1 and i_2:

$$i_{o1} = \beta i_1 - i_2 + I_A \tag{6.21}$$

where I_A is the bias current and β is the current gain of M_1-M_5 current mirror which can be expressed as [10]:

$$\beta = \frac{(1+\lambda Vds_{M4})}{(1+\lambda Vds_{M2})} \frac{(Vgs_{M4} - VTH_{M4})^2}{(Vgs_{M2} - VTH_{M2})^2} \frac{\left(W_4/L_4\right)}{\left(W_2/L_2\right)} \tag{6.22}$$

Assuming perfect matching between M_2 and M_4, β is equal to unity. The first input current i_1 applied to the flipped voltage follower based current mirror made of transistors M_1-M_2 and M_4, is transferred to the input terminal named node B where it is subtracted from second input current i_2. Negative feedback loop established by M_3

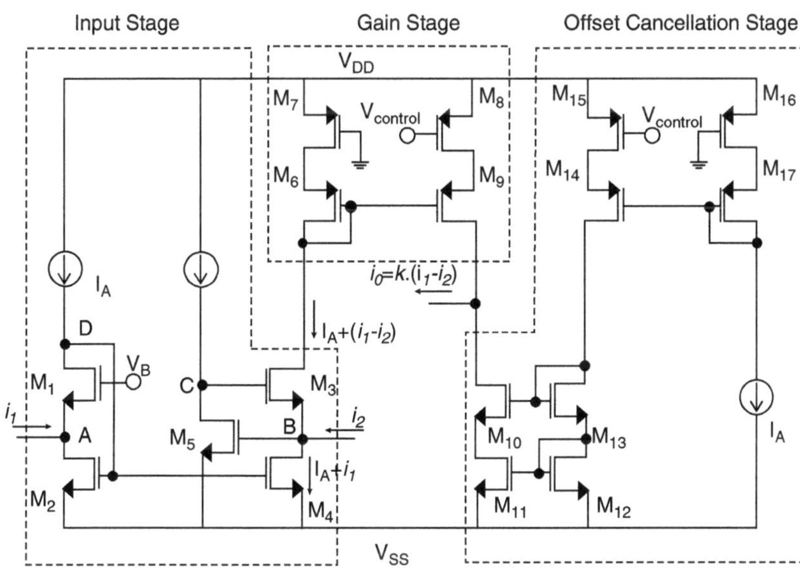

Fig. 6.12 Variable gain current mirror based CMIA where β is assumed unity [10]

and M_5 transistors provides a low input impedance at node B. Input impedance at node A is also very low, making the structure suitable for current inputs.

The second stage is a variable gain current mirror made of transistors M_6-M_9 that is used to amplify the output current from first stage and transfer it to output node. The current gain (K) of the M_6-M_9 current mirror is a function of control voltage $V_{control}$ as:

$$K = \frac{\frac{1}{gm_6} + r_{DS7}}{\frac{1}{gm_9} + r_{DS8}} \tag{6.23}$$

where:

$$r_{DS7} = \frac{1}{\mu C_{ox}\left(\frac{W}{L}\right)_7 \left(V_{DD} - \left|V_{THp}\right|\right)} \tag{6.24}$$

$$r_{DS8} = \frac{1}{\mu C_{ox}\left(\frac{W}{L}\right)_8 \left(V_{DD} - V_{control} - \left|V_{THp}\right|\right)} \tag{6.25}$$

According to Eqs. (6.23, 6.24 and 6.25), by adopting different choices for $V_{control}$, the values of r_{DS8} and consequently that of g_{m9} changes so resulting in an electronically variable gain for the CMIA of Fig. 6.12.

By changing $V_{control}$, the value of bias current at output branch changes. Therefore, a similar variable gain current mirror is used to produce the same current at the lower side of output branch to cancel any offset current at output node.

The output current is expressed as:

$$i_{o1} = k\left(\beta i_1 - i_2\right) \tag{6.26}$$

CMRR of this circuit is determined by input stage and the current gain of M_1-M_5 current mirror (see Eq. (6.22)), as follows:

$$CMRR = \frac{1}{\beta - 1} \tag{6.27}$$

6.1.5 CMIA Based on Electronically Current Gain Controlled Second Generation Current Conveyor

Electronically Current Gain Controlled Second-Generation Current Conveyor (ECCII) is another electronically tunable current-mode building block which was used to implement an electronically controllable CMIA in [11]. The symbolic representation of ECCII is shown in Fig. 6.13 and its operation is specified by Eq. (6.28):

$$V_x = V_y, \quad i_z = Ki_x, \ i_Y = 0 \tag{6.28}$$

In this device, the current applied at X terminal is amplified and transferred to node Z. Figure 6.14 shows a simple implementation of ECCII where current gain between X and Z terminals is varied by the voltage V_C, applied to variable gain current mirror, made of M_3-M_6 transistors.

Figure 6.15 shows the schematic of the ECCII-based CMIA of [11]. This circuit consists of only two ECCIIs and provides low input and high output impedances making it suitable for current-input current-output applications. The output of second ECCII is connected to the input of first ECCII which allows subtraction of common-mode currents and addition of differential-mode ones right at the first ECCII input. Performing a KCL analysis in Fig. 6.15 gives the output current as:

$$I_o = K_1 \left(I_1 - K_2 I_2\right) \tag{6.29}$$

where K_1 and K_2 are the current gains of the first and second ECCII respectively. For common-mode inputs of $I_1 = I_2 = I_{cm}$, common-mode gain CMIA is found as:

$$A_{cm} = \frac{I_{oc}}{I_{cm}} = K_1 \left(1 - K_2\right) \tag{6.30}$$

From (6.30), it comes that the current gain of second ECCII must be set close to unity:

$$K_2 = 1 - \varepsilon_i \tag{6.31}$$

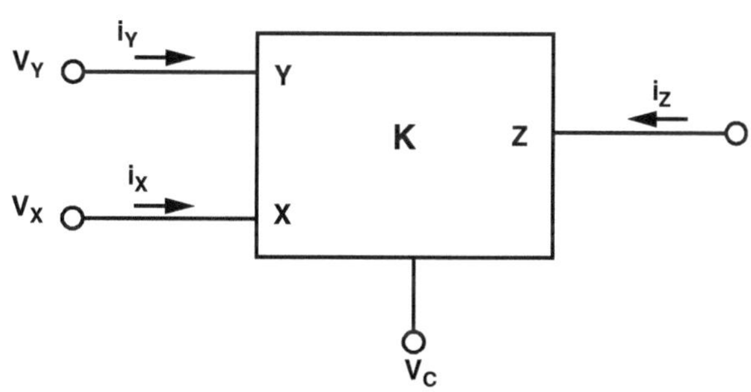

Fig. 6.13 Symbolic representation of ECCII [11]

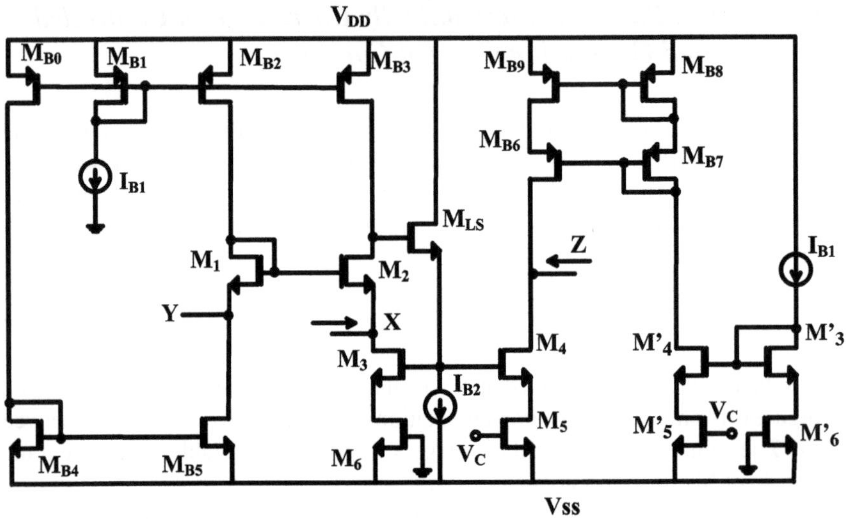

Fig. 6.14 ECCII circuit implementation [11]

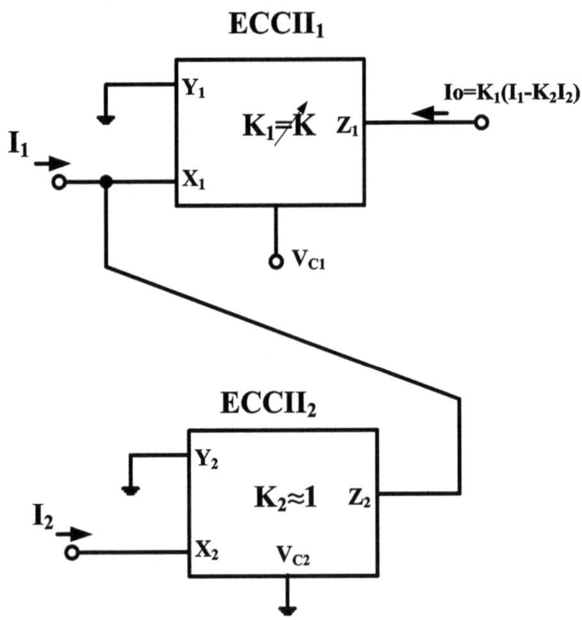

Fig. 6.15 Structure of the ECCII-Based CMIA in [11]

in which $\varepsilon_i \ll 1$ is the error term. By inserting Eq. (6.31) into Eq. (6.30), common-mode gain is found as:

$$A_{cm} = \frac{I_{oc}}{I_{cm}} = K_1 \varepsilon_i \tag{6.32}$$

For differential-mode inputs of $I_1 = -I_2 = I_{dm}/2$, differential-mode gain is found as:

$$A_{dm} = \frac{I_{od}}{I_{dm}} = K_1 \left(\frac{1}{2} + \frac{K_2}{2} \right) \tag{6.33}$$

Inserting Eq. (6.31) into Eq. (6.33) gives:

$$\mathrm{A_{dm}} = \mathrm{K_1} \left(1 - \frac{\varepsilon_i}{2} \right) \approx \mathrm{K_1} \tag{6.34}$$

Using Eqs. (6.34) and (6.32), CMRR is achieved as:

$$\mathrm{CMRR} = \frac{\mathrm{A_d}}{\mathrm{A_{cm}}} \cong \frac{1}{\varepsilon_i} \tag{6.35}$$

Compared to other CMIAs, circuit of Fig. 6.15 does not require matching between $\mathrm{ECCII_1}$ and $\mathrm{ECCII_2}$ to achieve high CMRR. Instead, CMRR is determined by current gain of $\mathrm{ECCII_2}$ which should be close to unity and if it deviates from unity due to unavoidable mismatches between transistors, its control voltage can be used to set current gain at unity.

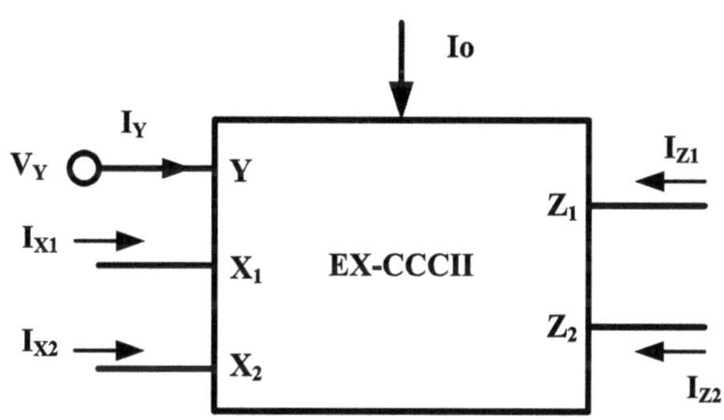

Fig. 6.16 Schematic representation of EX-CCCII [12]

6.1.6 CMIA Based on Extra X Current Controlled Current Conveyor (EX-CCCII)

In [12] a CMIA topology based on current controlled second-generation current conveyor with two X ports called EX-CCCII are introduced. In the reported CMIA, input signals are currents while output signals are in either current or voltages. The schematic representation of EX-CCCII is shown in Fig. 6.16. The relation between current and voltage signals of its ports is represented as:

$$\begin{cases} I_Y = 0 \\ V_{X1} = I_{X1} r_{X1} + \alpha_1 V_Y \\ V_{X2} = I_{X2} r_{X2} + \alpha_2 V_Y \\ I_{Z1} = \beta_1 I_{X1} \\ I_{Z2} = \beta_2 I_{X2} \end{cases} \tag{6.36}$$

In (6.36), α_1 and α_2 are the voltage tracking gains between Y and X_1 ports and Y and X_2 ports, respectively, while β_1 and β_2 are current tracking gains between X_1 and Z_1 ports and X_2 and Z_2 ports respectively. Moreover, r_{x1} and r_{x2} are intrinsic resistances at X_1 and X_2 ports respectively and their values can be electronically controlled by I_o current source.

Figure 6.17 shows a simple implementation of EX-CCCII reported in [2]. In this circuit r_{X1} and r_{X2} can be controlled by I_o current source according to Eq. (6.37) with usual meaning of symbols [12]:

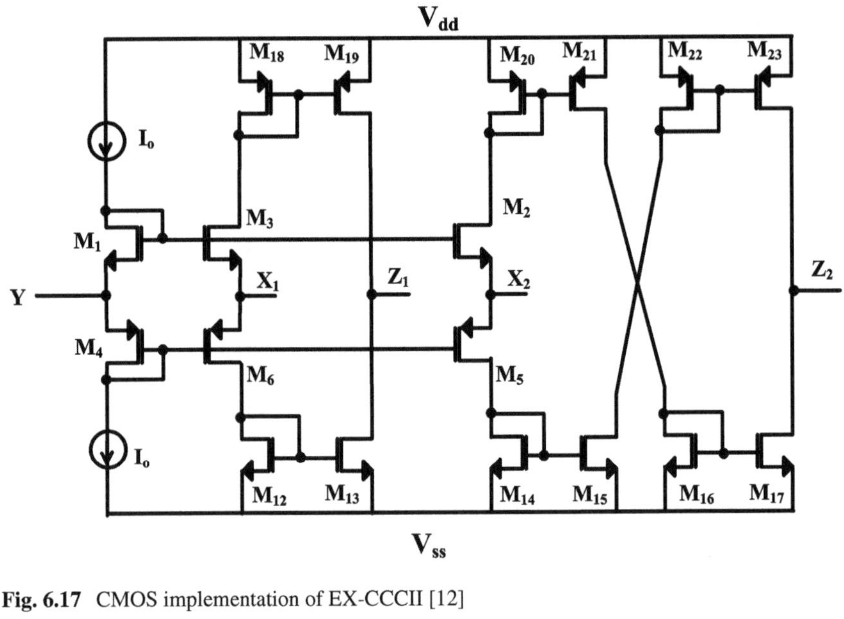

Fig. 6.17 CMOS implementation of EX-CCCII [12]

$$r_{X1} = r_{X2} = \frac{1}{\sqrt{8\mu C_{ox}\left(\frac{W}{L}\right)I_o}} \tag{6.37}$$

Here W and L are channel wide and length of transistors at X_1 (M_3, M_6) and X_2 (M_2, M_5) inputs. In this circuit, the current tracking gain between X_2 and Z_2 terminals is negative.

The CMIA topology of [12] is shown in Fig. 6.18. In this circuit, input currents I_{in1} and I_{in2} are applied to X_1 and X_2 ports of EX-CCCII$_1$ while its Z_1 and Z_2 ports are tied together to make a current subtraction node. Therefore, difference between I_{in1} and I_{in2} is produced and applied to the Y port of EX-CCCII$_2$. The X_1 port of EX-CCCII is terminated to R_1 and its Z_1 output is connected to its Y port. Consequently, from Y port, EX-CCCII$_2$ operated as resistor simulator equal to:

$$R_{Y2} = \frac{R_1 + r'_{X1}}{\alpha'_1 \beta'_1} \tag{6.38}$$

where r'_{X1}, α'_1 and β'_1 denoted the related parameters of EX-CCCII$_2$. The difference between I_{in1} and I_{in2} is converted to a voltage by R_{Y2} producing V_{out} and I_{out} as follows:

$$V_{out} = \left(\beta_1 I_{in1} - \beta_2 I_{in2}\right)\frac{R_1 + r'_{X1}}{\alpha'_1 \beta'_1} \tag{6.39}$$

$$I_{out} = \left(\beta_1 I_{in1} - \beta_2 I_{in2}\right)\frac{R_1 + r'_{X1}}{\alpha'_1 \beta'_1 r'_{X2}} \tag{6.40}$$

Here the A_{dm}, A_{cm} and CMRR relations for V_{out} are derived. The similar analysis can be performed for I_{out}. For differential-mode inputs of $I_{in1} = -I_{in2} = I_{dm}/2$, using Eqs. (6.39) and (6.40), A_{dm} is found as:

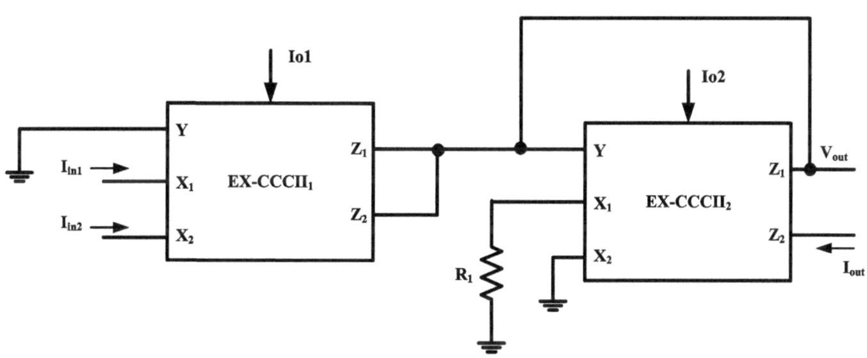

Fig. 6.18 EX-CCCII based CMIA of [12]

$$A_{dm} = \frac{V_{out}}{I_{dm}} = (\beta_1 + \beta_2)\frac{R_1 + r'_{X1}}{2\alpha'_1 \beta'_1} \quad (6.41)$$

According to Eq. (6.41), A_{dm} can be controlled by varying r'_{X1}.

For common-mode inputs of $I_{in1} = I_{in2} = I_{cm}$, using Eq. (6.39), A_{cm} is found as:

$$A_{cm} = \frac{V_{out}}{I_{cm}} = (\beta_1 - \beta_2)\frac{R_1 + r'_{X1}}{\alpha'_1 \beta'_1} \quad (6.42)$$

Using Eqs. (6.41) and (6.42), CMRR is found as:

$$CMRR = \frac{\beta_1 + \beta_2}{2(\beta_1 - \beta_2)} \quad (6.43)$$

As Eq. (6.43) indicates, CMRR is solely determined by the matching between current gain parameters of the first EX-CCCII. This circuit requires a voltage buffer at V_{out} output.

6.1.7 CMIA Based on Current Differencing Trans-Resistance Amplifier (CDTRA)

Figure 6.19 shows a CMIA topology based on Current Differencing Trans-Resistance Amplifier (CDTRA), presented in [13]. The input signals are applied to *p* and *n* terminals of CDTRA which are low impedance current input ports. A voltage proportional to the difference between input currents is produced at port X equal to:

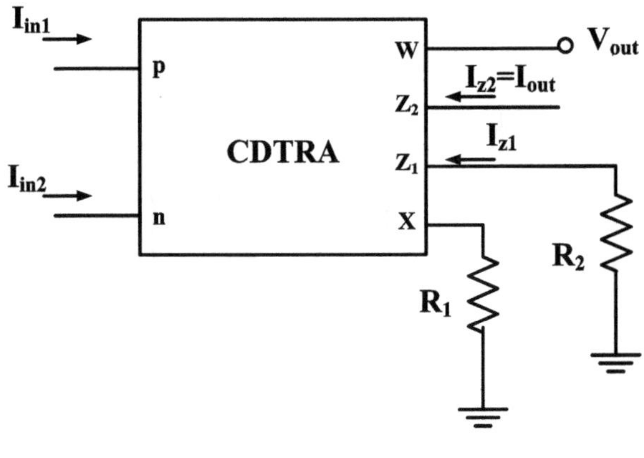

Fig. 6.19 CDTRA based CMIA of [13]

$$V_X = r_m\left(K_1 I_{in1} - K_2 I_{in2}\right) \tag{6.44}$$

The parameter r_m is the trans-resistance gain which is electronically controllable and K_1 and K_2 are current gains close to unity. There is current tracking between X-Z_1 and X-Z_2 terminals by current gains of β_1 and β_2 respectively with close to unity values. The voltage at Z_2 terminal is also conveyed to W port by a gain of α_w. In the CMIA of Fig. 6.19, the voltage at X port is converted to a proportional current by R_1 and conveyed to Z_1 and Z_2 ports by current gains of β_1 and β_2 respectively. Therefore, for I_x we have:

$$I_x = \frac{r_m\left(K_1 I_{in1} - K_2 I_{in2}\right)}{R_1} \tag{6.45}$$

The current at X port is conveyed to Z_2 port by gain of β_2 producing output current as:

$$I_{out} = \beta_2 \frac{r_m\left(K_1 I_{in1} - K_2 I_{in2}\right)}{R_1} \tag{6.46}$$

For differential-mode current of $I_{in1} = -I_{in2} = I_{dm}/2$, A_{dm} is found as:

$$A_{dm} = \frac{I_{out}}{I_{dm}} = \beta_2 \frac{r_m}{2R_1}\left(K_1 + K_2\right) \tag{6.47}$$

According to Eq. (6.47), differential-mode gain can be electronically controlled by r_m. For common-mode inputs of $I_{in1} = I_{in2} = I_{cm}$, A_{cm} is found as:

$$A_{cm} = \frac{I_{out}}{I_{cm}} = \beta_2 \frac{r_m}{R_1}\left(K_1 - K_2\right) \tag{6.48}$$

As it is seen from Eq. (6.47), common-mode gain is determined by matching between α_1 and α_2 parameters.

From Eqs. (6.47) and (6.48), CMRR is found as:

$$CMRR = \frac{K_1 + K_2}{2\left(K_1 - K_2\right)} \tag{6.49}$$

In the CMIA of Fig. 6.19, the resistor R_2 connected to Z terminal converts the Z port current to a proportional voltage which is transferred to W output:

$$V_{out} = \alpha_w \beta_2 \frac{r_m\left(K_1 I_{in1} - K_2 I_{in2}\right)}{R_1} R_2 \tag{6.50}$$

In this equation α_w has a value close to unity and is the voltage transfer gain between Z_2 and W ports. For CMOS implementation of CDTRA the reader is referred to [13].

6.1.8 CMIA Based on Current Follower Differential Input Transconductance Amplifier

In [14] an electronically controllable CMIA based on a recently introduced current-mode active building block called current follower differential input transconductance amplifier (CFDITA) is reported. This device comprises a current follower stage followed by two operational transconductance amplifiers (OTA) stages which produce two current with opposite phase. First OTA produces I_{out1} and second OTA produces I_{out2}. Figure 6.20 shows the symbolic representation of CFDITA and the CMIA structure. It is a five terminal device and its g_{m1} and g_{m2} characteristics can be controlled by current sources I_{B1} and I_{B2}, respectively. The relation between its terminals currents and voltages are shown by following matrix:

$$\begin{bmatrix} I_V \\ I_Z \\ I_{O+} \\ I_{O-} \end{bmatrix} = \begin{bmatrix} 0 & 0 & 0 & 0 & 0 \\ \beta_1 & 0 & 0 & 0 & 0 \\ 0 & \gamma_1 g_{m1} & -\gamma_1 g_{m1} & 0 & 0 \\ 0 & -\gamma_2 g_{m2} & \gamma_2 g_{m2} & 0 & 0 \end{bmatrix} \begin{bmatrix} I_F \\ V_Z \\ V_V \\ V_{O+} \\ V_{O-} \end{bmatrix} \tag{6.51}$$

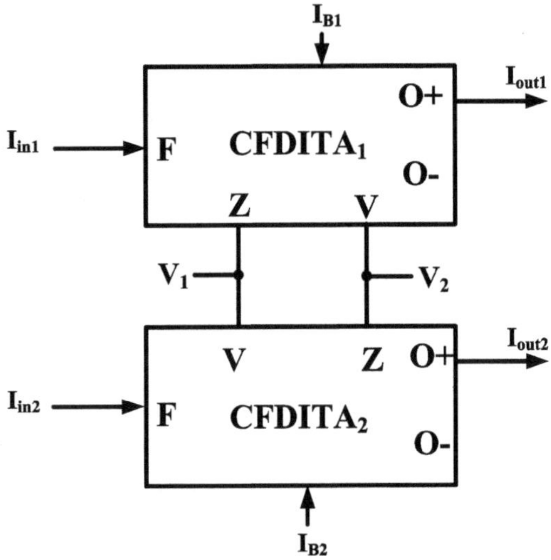

Fig. 6.20 CFDITA-based CMIA of [14]

where β_1 is the current transfer gain from F to Z terminals. Parameters γ_1 and γ_2 are transconductance inaccuracies between $V_Z - V_V$ to I_{O+} and $V_V - V_Z$ to I_{O-}, respectively [14].

In the CMIA of Fig. 6.20, the Z terminal of each active building block is connected to the V of other device. The input currents applied to F terminals are conveyed to Z terminals where they are converted to a proportional voltage equal to:

$$V_1 = \beta_1 I_{in1} \left(R_{Z1} \parallel R_{V2}\right) \tag{6.52}$$

$$V_2 = \beta_2 I_{in2} \left(R_{Z2} \parallel R_{V1}\right) \tag{6.53}$$

Where β_i, R_{Zi} and R_{Vi} for i = 1–2 are current transfer gain, Z terminal impedance and V terminal impedance of the related device respectively. Referring to Eq. (6.51), the difference between V_1 and V_2 is converted to output currents expressed as:

$$I_{out1} = \gamma_1 g_{m1} V_1 - \gamma_1 g_{m1} V_2 \tag{6.54}$$

$$I_{out2} = \gamma_2 g_{m2} V_2 - \gamma_2 g_{m2} V_1 \tag{6.55}$$

Inserting Eqs. (6.52 and 6.53) into Eqs. (6.54 and 6.55), it results:

$$I_{out1} = \gamma_1 g_{m1} \left[\beta_1 I_{in1} \left(R_{Z1} \parallel R_{V2}\right) - \beta_2 I_{in2} \left(R_{Z2} \parallel R_{V1}\right)\right] \tag{6.56}$$

For common-mode input currents of $I_{in1} = I_{in2} = I_{cm}$, from Eq. (6.56) A_{cm} is achieved as:

$$A_{cm} = \frac{I_{out1}}{I_{cm}} = \gamma_1 g_{m1} \left[\beta_1 \left(R_{Z1} \parallel R_{V2}\right) - \beta_2 \left(R_{Z2} \parallel R_{V1}\right)\right] \tag{6.57}$$

For differential-mode input currents of $I_{in1} = -I_{in2} = I_{dm}/2$, from Eq. (6.56) A_{dm} is achieved as:

$$A_{dm} = \frac{I_{out1}}{I_{dm}} = \frac{1}{2} \gamma_1 g_{m1} \left[\beta_1 \left(R_{Z1} \parallel R_{V2}\right) + \beta_2 \left(R_{Z2} \parallel R_{V1}\right)\right] \tag{6.58}$$

The value of A_{dm} can be controlled by g_{m1} through current source I_{B1} according to:

$$g_m = \sqrt{\mu C_{0x} \frac{W}{L} I_{B1}} \tag{6.59}$$

where μ, C_{ox}, W and L are physical parameters of the MOS transistors used in the first OTA stage. From Eqs. (6.57 and 6.58), CMRR for I_{out1} is found as:

$$CMRR = \frac{\beta_1 (R_{Z1} \parallel R_{V2}) + \beta_2 (R_{Z2} \parallel R_{V1})}{2\left[\beta_1 (R_{Z1} \parallel R_{V2}) - \beta_2 (R_{Z2} \parallel R_{V1})\right]} \tag{6.60}$$

Similarly, I_{out2} output can be controlled by g_{m2} and I_{B2}.

The value of A_{dm} and CMRR can be increased by connecting O+ output of each device to O− of other device [14].

6.2 CMIA Based on Electronically Controllable Resistors

6.2.1 Electronically Controllable CMIA Using a Single MOS Transistor Operating in Triode Region as a Variable Resistor

In the CMIA presented [15], a nMOS transistor operating in triode region is used to electronically change the gain. Two single-input multiple output current operational amplifiers (SI-MO COA) in negative feedback configuration are utilized as active building blocks. The circuit has a current-input current-output topology. Symbolic representation of SI-MO COA is shown in Fig. 6.21 which exhibits ideally zero input resistance ($r_{in} = 0$), infinite open loop current gain ($A_0 = \infty$) and infinite output resistance ($r_o = \infty$). Figure 6.22 shows the CMIA structure where two SI-MO COAs along with resistors R_1-R_2 are employed in negative feedback configuration. The cross connection of the outputs COA_1 and COA_2 makes a current subtraction node

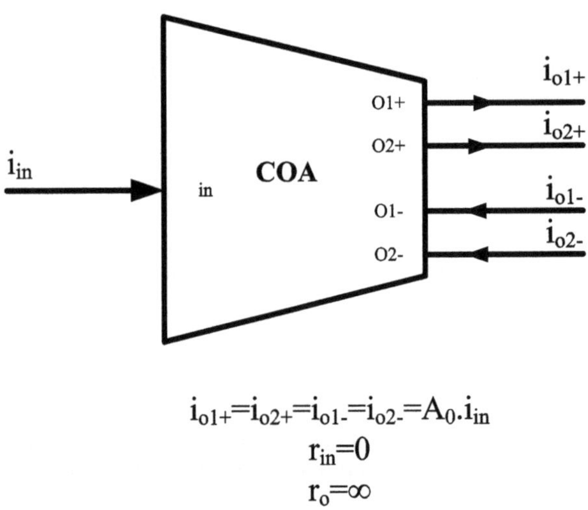

Fig. 6.21 SI-MO COA symbolic representation [15]

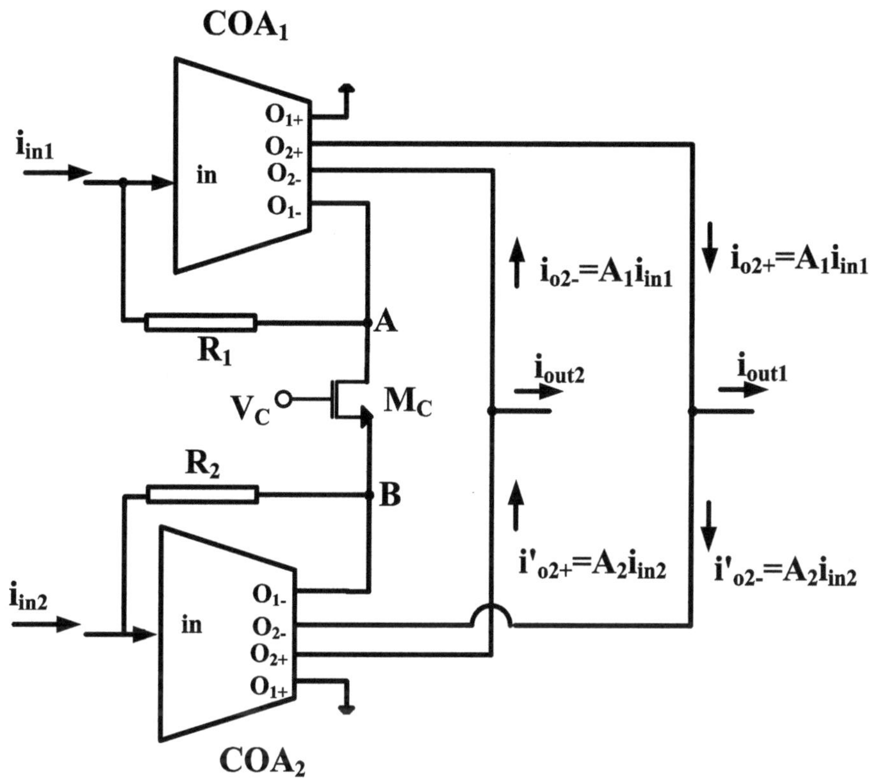

Fig. 6.22 SI-MO COA based CMIA of [15]

to attenuate common-mode inputs and amplify differential-mode ones. Transistor M_C operating in triode region is used as a variable resistor.

In Fig. 6.22, KCL analysis at output nodes gives the output currents of CMIA as:

$$i_{out1} = i_{o2+} - i'_{o2-} = A_1 i_{in1} - A_2 i_{in2} \tag{6.61}$$

$$i_{out2} = i'_{o2+} - i_{o2-} = A_2 i_{in2} - A_1 i_{in1} \tag{6.62}$$

where A_1 and A_2 denote the closed-loop current gains of COA_1 and COA_2 respectively. However, as the circuit behaves differently for common-mode and differential-mode inputs, A_1 and A_2 should be calculated separately for each mode.

For common-mode inputs, by the assumption that $R_1 = R_2 = R$, nodes A and B have equal voltages. Therefore, no current flows into transistor M_C that can be considered as open circuit. By assuming matching between related feedback resistors, the equivalent circuit in common-mode is shown in Fig. 6.23 where both COAs perform as current buffers for common-mode inputs of $i_{in1} = i_{in2} = i_{cm}$ with gains close to unity as expressed by:

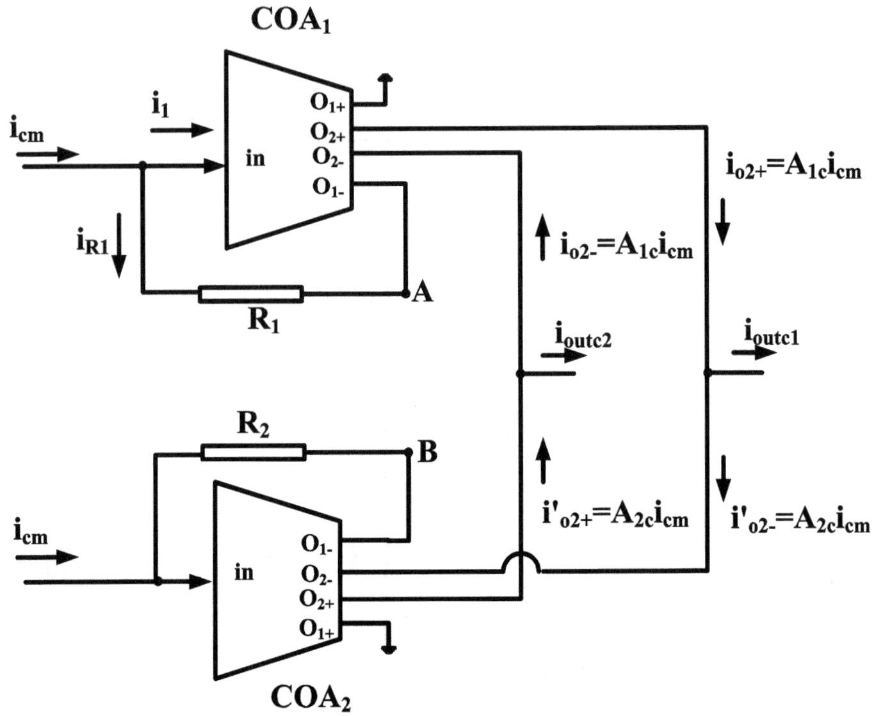

Fig. 6.23 Common-mode equivalent circuit of CMIA of [15]

$$A_{c1} = \frac{A_{01}}{1 + A_{01}} \approx 1 \tag{6.63}$$

$$A_{c2} = \frac{A_{02}}{1 + A_{02}} \approx 1 \tag{6.64}$$

where A_{0i} and A_{ci} for $i = 1$–2 are the open loop and closed loop gains of related COA respectively. Using Eqs. (6.63) and (6.64), the common-mode output signals of CMIA are expressed as:

$$i_{outc1} = \left(A_{c1} - A_{c2}\right) i_{cm} \tag{6.65}$$

$$i_{outc2} = -\left(A_{c1} - A_{c2}\right) i_{cm} \tag{6.66}$$

where i_{outc1} and i_{outc2} are the common-mode output currents of COA_1 and COA_2 respectively.

By assuming a mismatch of $\Delta A \ll 1$ between Ac_1 and Ac_2 and defining $A_{c1} = 1 - \Delta A$ and $A_{c2} = 1 + \Delta A$, using Eqs. (6.65) and (6.66), the common-mode gain of CMIA is found as:

$$|A_{cm}| = \left|\frac{i_{outc1}}{i_{cm}}\right| = \left|\frac{i_{outc2}}{i_{cm}}\right| = 2\Delta A \tag{6.67}$$

It can be shown that the negative feedback configuration reduces the sensitivity of Ac to mismatches between open loop current gains of COA_1 and COA_2 respectively. For example, by assuming $A_{01} = 1100$ and $A_{02} = 900$, using Eqs. (6.65) and (6.66), A_{c1} and A_{c2} are calculated as 0.999 and 0.9988 giving a ΔA of 0.0002, so resulting −68dB common-mode gain for CMIA. The negative feedback loop also reduces the input impedence making the CMIA fully appropriate for current input signals.

In Fig. 6.21, the output currents for differential-mode inputs of $i_{in1} = -i_{in2} = i_{dm}/2$ are:

$$i_{outd1} = i_{o2+} - i'_{o2-} = (A_{d1} + A_{d2})i_{dm}/2 \tag{6.68}$$

$$i_{outd2} = i'_{o2+} - i_{o2-} = -(A_{d2} + A_{d1})i_{dm}/2 \tag{6.69}$$

being A_{d1} and A_{d1} the differential-mode closed loop gains of COA_1 and COA_2 respectively. In differential-mode, by assuming $R_1 = R_2 = R$, we have $V_A = -V_B$, and a straight forward analysis gives A_{d1} and A_{d2} as:

$$|A_{d1}| = |A_{d2}| = \left|\frac{i_{o2+}}{i_{dm}/2}\right| = \left|\frac{i_{o2-}}{i_{dm}/2}\right| = \left|\frac{i_{o1-}}{i_{dm}/2}\right| = \left|\frac{i'_{o2+}}{i_{dm}/2}\right| = \left|\frac{i'_{o1-}}{i_{dm}/2}\right| = \left|\frac{i'_{o2-}}{i_{dm}/2}\right| = \left(1 + \frac{2R}{R_C}\right) \tag{6.70}$$

where R_C is the equivalent resistance of M_C.

Inserting Eq. (6.70) into Eqs. (6.68 and 6.69), the overall differential-mode gain is found as:

$$A_{dm} = \frac{i_{outd1,2}}{i_{dm}} = \left(1 + \frac{2R}{R_C}\right) \tag{6.71}$$

Using Eqs. (6.71) and (6.67), the CMRR is given as:

$$CMRR = \left(1 + \frac{2R}{R_C}\right)\frac{1}{2\Delta A} \tag{6.72}$$

Here CMRR analysis is performed by assuming matched resistors. In order to take into account the mismatches between resistors, the analysis performed for 3-Op-Amp based IA in Chap. 1 based on superposition theorem should be used for the CMIA of Fig. 6.22.

A possible CMOS implementaion of SI-MO COA presented in [15] is shown in Fig. 6.24.

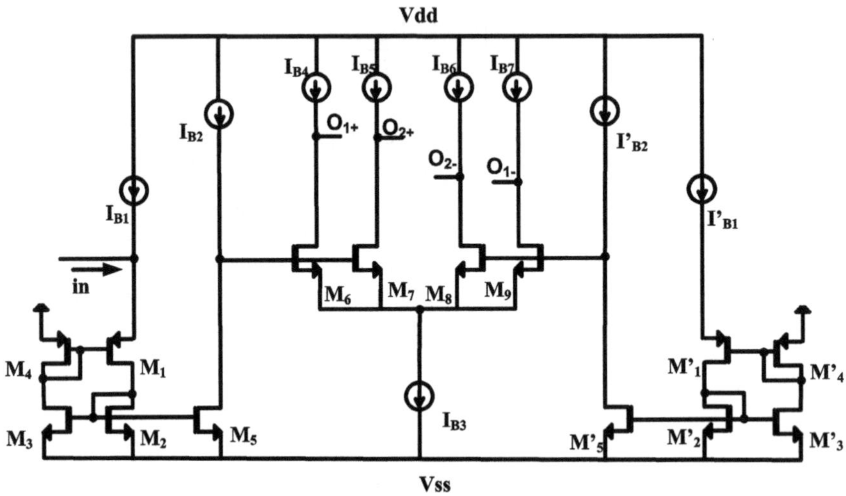

Fig. 6.24 A simple implementation of SI-MO COA reported in [15]

6.2.2 Electronically Controllable CMIA Using Two MOS Transistors Operating in Saturation Region as a Variable Resistor

In [16], a CMIA using eight MOS transistors forming a current mirror, a variable resistor and an inverting amplifier is reported. The circuit is shown in Fig. 6.25 in which the first stage is a class AB current mirror formed by transistors M_1-M_4. The second input, I_2, is directly connected to the output of M_1-M_4 current mirror to make a current subtraction node. The second stage is simply an inverting amplifier which transfers the voltage across active resistor R_1 (made of transistors M_{R1}-M_{R2}) to the output node. Here, to provide low output impedance, large aspect ratios must be chosen for transistors M_5 and M_6. A simple analysis of the circuit of Fig. 6.25 yields the output voltage as:

$$V_{out} = \alpha R_1 \left(\beta I_1 - I_2\right) \tag{6.73}$$

where β and α are non-ideal current gain of the current mirror M_1-M_4 and voltage gain of inverting voltage amplifier M_5-M_6, respectively. Differential-mode gain (for $I_1 = -I_2 = I_{dm}/2$), common-mode gain (for $I_1 = I_2 = I_{cm}$) and and CMRR of the CMIA of Fig. 6.25 are achieved using Eq. (6.73) as:

$$A_{dm} = \frac{V_{outd}}{I_{dm}} = \frac{1}{2} R_1 \alpha \left(\beta + 1\right) \tag{6.74}$$

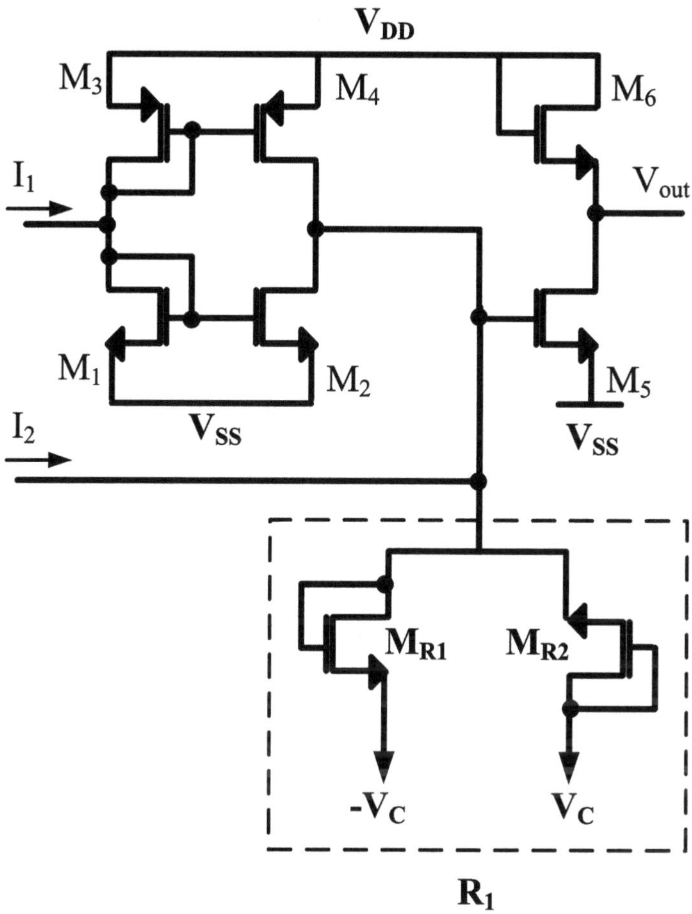

Fig. 6.25 The current mirror based CMIA of [16]

$$A_{cm} = \frac{V_{outc}}{I_{cm}} = R_1\alpha\left(\beta - 1\right) \tag{6.75}$$

$$CMRR = \frac{A_{dm}}{A_{cm}} = \frac{\left(\beta + 1\right)}{2\left(\beta - 1\right)} \tag{6.76}$$

According to Eq. (6.74), the value of A_d can be controlled by R_1. In Fig. 6.25 the dashed path shows the realization of R_1 by two MOS transistors; its value can be adjusted by control voltages V_c and $-V_c$ according to:

$$R_1 = \frac{L}{2\mu C_{ox} W\left(V_C - V_{TH}\right)} \tag{6.77}$$

The same variable resistor is used in [17] to design an electronically controllable CMIA based on modified Z copy current differencing trans-conductance amplifier (MZC-CDTA).

6.2.3 Electronically Controllable CMIA Using Two MOS Transistors Operating in Triode Region as Variable Resistor

In [18], Operational Trans-Resistance Amplifier (OTRA) is used for implementing a linear variable resistor which is used to electronically control the differential mode gain of CMIA. Figure 6.26 shows the concept of loperation in which two NMOS transistors M_1 and M_2 are matched and operating in the ohmic region. The current of an MOS transistor operating in ohmic region is expressed as [18]:

$$I = \mu C_0 \frac{W}{L}\left(V_G - V_{TH}\right)\left(V_D - V_S\right) + a_1\left(V_D^2 - V_S^2\right) + a_2\left(V_D^3 - V_S^3\right) \tag{6.78}$$

where V_G, V_D, V_S and V_{TH} are the gate, drain, source and threshold voltages of transistor, respectively. Parameters a_1 and a_2 are constant coefficients. In Fig. 6.26 since both transistors have equal drain and source voltages, if we produce $I_1 - I_2$ both the even and odd non-linearities are cancelled and a linear function is obtained as:

$$I_1 - I_2 = \mu C_0 \frac{W}{L}\left(V_a - V_b\right)\left(V_i - V_2\right) = G\left(V_i - V_2\right) \tag{6.79}$$

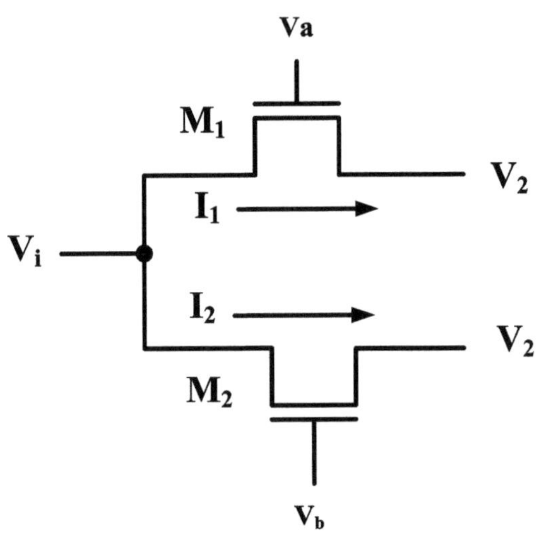

Fig. 6.26 Variable resistor implemented by two MOS transistors in ohmic region [18]

In Eq. (6.79) the value of G can be controlled by V_a and V_b.

Various methods can be employed to subtract I_1 and I_2. A suitable active element for this purpose is the OTRA because it has two current input terminals which are virtually grounded [18]. The operation matrix of OTRA is [18, 19]:

$$\begin{bmatrix} V_p \\ V_n \\ V_o \end{bmatrix} = \begin{bmatrix} 0 & 0 & 0 \\ 0 & 0 & 0 \\ R_m & -R_m & 0 \end{bmatrix} \begin{bmatrix} I_p \\ I_n \\ I_o \end{bmatrix} \tag{6.80}$$

Figure 6.27 shows the implementation of a linear resistor connected between negative input and output terminals of OTRA [18]. Here the main source of error is the finite transconductance gain of OTRA.

Figure 6.28 shows the electronically controllable CMIA which utilizes the above mentioned resistor. In Fig. 6.28a, the differential gain is [18]:

$$A_d = \frac{V_o}{i_2 - i_1} = \frac{R_{\text{var}} R_1}{R_2} \tag{6.81}$$

Figure 6.28b shows its MOS-based implementation. The common-mode gain and CMRR are determined by the matching between the two resistors R_1 in first stage, by that between two resistors R_2 in the output stage and also by the matching between parasitic capacitances of $OTRA_1$ and $OTRA_2$.

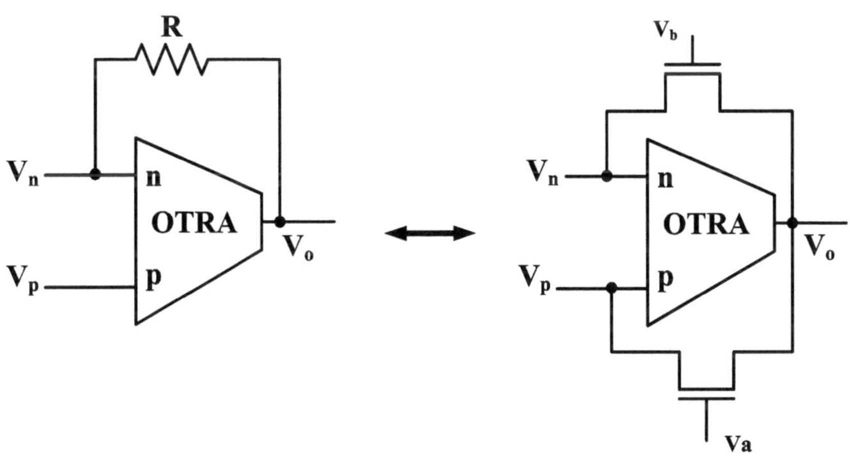

Fig. 6.27 Implementation of a linear variable resistance connected between negative input and output terminals of OTRA [18]

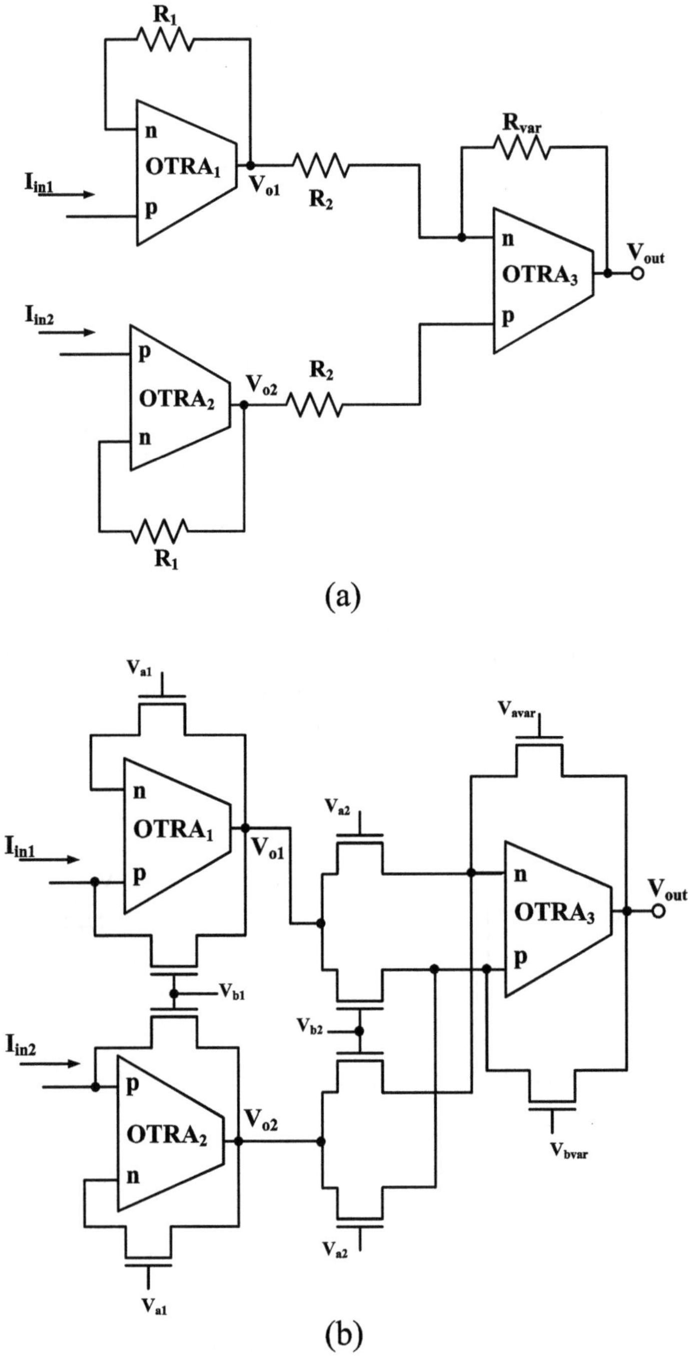

Fig. 6.28 OTRA based CMIA of [18] (**a**) resistor-based implementation (**b**) MOS based implementation

References

1. Jaikla W., Siripruchyanan M. (2006) Dual-outputs current controlled differential voltage current conveyor and its applications.International Symposium on Communications and Information Technologies, 2006.
2. Siripruchyanun M., Jaikla W. (2008) Current controlled current conveyor trans-conductance amplifier (CCCCTA): A building block for analog signal processing. Electronics Engineering, 90: 443–453.
3. Chanapromma C., Tanaphatsiri C., Siripruchyanun M. (2009) An electronically controllable instrumentation amplifier based on CCCCTAs. International Symposium on Intelligent Signal Processing and Communications Systems, 2009.
4. Siripruchyanun M., Payakkakul K. (2017) A temperature-insensitive instrumentation amplifier using CCTA-based voltage to current converter. International Electrical Engineering Congress (iEECON), 2017.
5. Fabre A., Saaid O., Wiest F., Boucheron C. (1996) High frequency applications based on a new current controlled conveyor. IEEE Transactions on Circuits and Systems I: Fundamental Theory and Applications, 43(2):81–92.
6. Maheshwari S. (2002) High CMRR wide bandwidth instrumentation amplifier using current controlled conveyors.International Journal of Electronics, 89(12):889–896.
7. H. Ercan, S. A. Tekin, M. Alci, "Voltage- and Current-Controlled High CMRR Instrumentation Amplifier Using CMOS Current Conveyors," Turkish Journal of Electrical Engineering and Computer Sciences, Vol. 20, No. 4, pp. 547–556, 2012.
8. Harzi Z. M., Alami M. (2015) A novel high bandwidth current mode instrumentation amplifier. International Conference on Microelectronics (ICM), 2015.
9. Abuelmaatti M. T., Al-Qahtani M. A. (1998) A new current-controlled multiphase sinusoidal oscillator using translinear current conveyors. IEEE Transactions on Circuits and Systems II: Analog and Digital Signal Processing, 45(7):881–885.
10. Safari L., Minaei S. (2013) A novel resistor-free electronically adjustable current-mode instrumentation amplifier. Circuits Systems and Signal Processing, 32:1025–1038.
11. Safari L., Minaei S. (2017) New ECCII-Based electronically controllable current-mode instrumentation Amplifier with High Frequency Performance. European Conference on Circuit Theory and Design (ECCTD), 2017.
12. Agrawal D., Maheshwari S. (2018) Cascadable current mode instrumentation amplifier. AEÜ - International Journal of Electronics and Communications, 92:91–101, 2018.
13. Ayten U., Cem Dikbaş M. (2018) Current and tansimpedance mode instrumentation amplifier using a single new active component named CDTRA.AEÜ - International Journal of Electronics and Communications, 91:24–36, 2018.
14. Chaturvedi B., Kumar A. (2018) Electronically tunable current-mode instrumentation amplifier with high CMRR and Wide bandwidth. AEU-International Journal of Electronics and Communications, 116–123.
15. Safari L., Minaei S. (2017) A novel COA-based electronically adjustable current-mode instrumentation amplifier topology. AEU-International Journal of Electronics and Communications, 82:285–293, 2017.
16. Safari L., Yuce E., Minaei S. (2016) A new transresistance-mode instrumentation amplifier with low number of MOS transistors and electronic tuning opportunity. Journal of Circuits, Systems, and Computers 25(4), 2016.
17. Oruganti S., Pandey N., Pandey R. (2018) Electronically tunable high gain current-mode instrumentation amplifier" AEÜ - International Journal of Electronics and Communications, in press, 2018.
18. Pandey R., Pandey N., Paul S. K. (2013) Electronically tunable transimpedance instrumentation amplifier based on OTRA.Journal of Engineering, 10:1–5.
19. Salama K. N., Soliman A. M. (1999) CMOS operational transresistance amplifier for analog signal processing. Microelectronics Journal, 30(3):235–245.

Chapter 7
Mismatch Implications in Current-Mode Instrumentation Amplifiers

7.1 Mismatch Overview in CMOS Technology

Two identical designed transistors on an integrated circuit have unavoidable random variations in their dimensions and show different performances. As an example, Fig. 7.1 shows the local variations occurring in length of two MOS transistors. Such variations cause the effective dimensions of transistors to be random parameters with variance of [1]:

$$\sigma_L^2 \propto \frac{1}{W} \tag{7.1}$$

$$\sigma_W^2 \propto \frac{1}{L} \tag{7.2}$$

By increasing the size of transistors, the variations in dimensions tend to decrease. According to Eqs. (7.1 and 7.2), there is better matching for larger size transistor dimensions.

As technology scales down, mismatches between devices become more severe. The reason is that the percent errors increase for smaller fabrication dimensions. For example, the increasing effect of dopant variation in fine technologies worsens the threshold voltage mismatch between adjacent transistors [2]. For a matched pair of MOS transistors, threshold voltage differences ΔV_{TH} and current factor differences $\Delta\beta$ ($\beta = \mu C_{OX}W/L$, being μ the mobility and C_{ox} the oxide capacitance) cause unequal drain source current and gate source voltage. These differences are modeled as two independent random parameters with normal distribution having zero mean value and area-dependent variance of [3]:

$$\sigma^2\left(\Delta V_{TH}\right) = \frac{A_{VTH}^2}{WL} \tag{7.3}$$

G. Ferri et al., *Current-Mode Instrumentation Amplifiers*, Analog Circuits and Signal Processing, https://doi.org/10.1007/978-3-030-01343-1_7

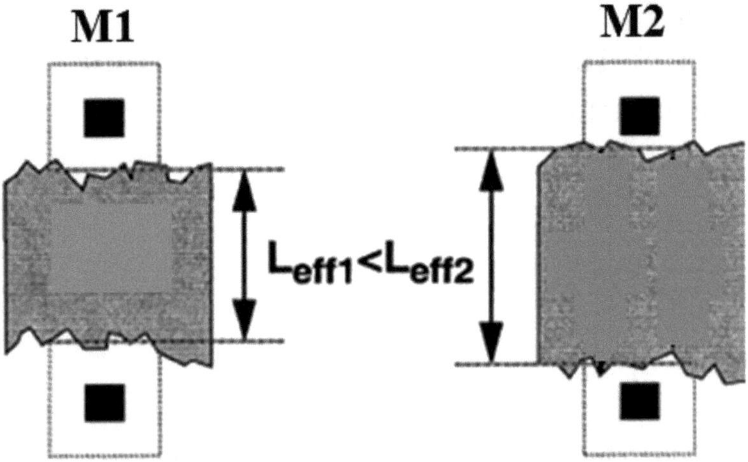

Fig. 7.1 Random variations in L of two matched MOS transistors [1]

$$\left(\frac{\sigma(\Delta\beta)}{\beta}\right)^2 = \frac{A_\beta^2}{WL} \tag{7.4}$$

In Eqs. (7.3 and 7.4), W is channel width, L is channel length and A_{VTH} and A_β are technology dependent area parameters. Figure 7.2 shows the A_{VTH} and A_β variations in different technologies. V_{TH} variation with respect to $1/(WL)$ for different technologies is also shown in Fig. 7.3. It shows that matching improves for larger size transistors. It is important to note that for the unequal body source voltage of transistors, the body effect worsens the V_{TH} mismatch.

7.2 Mismatch Effect on the CMRR in Various Structures of CMIAs

In order to study the effect of mismatches on the CMRR of CMIAs, let us consider the voltage input type (i.e., with voltage/current output) as the most popular topology of CMIAs shown in Figs. 7.4 and 7.5, respectively. Here, the CMRR is usually determined by the first stage. The second stage is used to produce the required output current/voltage signal and usually does not have any effect on the overall CMRR. If the output signal is a current, the second stage can be simply the load resistor while, for an output voltage, since Z shows a high impedance, a sort of voltage buffer is mandatory in second stage. The CMIA of Fig. 7.4 employs two active building blocks such as CCII, OFCC, COA etc. to transfer Y terminal voltage to X terminal. The transferred voltage across resistor R_1 is converted to a proportional current signal. This current is copied to output (Z terminal) by current transfer

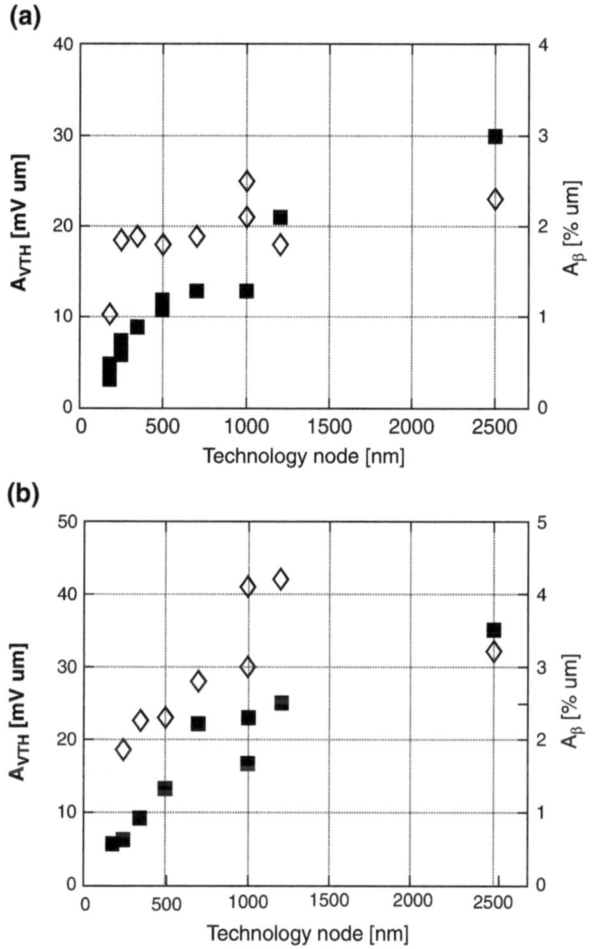

Fig. 7.2 Matching parameters A_{VTH} (■) and A_β(◇) for (**a**) NMOS (**b**) PMOS transistors [3]

function between X and Z terminals and converted to the appropriate output signal by second stage (or R_2). By ignoring the effect of non-zero impedances at X terminals, the CMRR of Fig. 7.4 is determined by the voltage transfer gains between Y_1 and Y_2 terminals of the used active elements as [5]:

$$CMRR = \frac{A_{dm}}{A_{cm}} = \frac{\beta_1 + \beta_2}{2(\beta_1 - \beta_2)} \tag{7.5}$$

where β_i for $i = 1, 2$ is the voltage transfer gain of the associated active building block and can be expressed in terms of voltage transfer error $\varepsilon_i << 1$ as:

$$\beta_i = 1 \pm \varepsilon_i \tag{7.6}$$

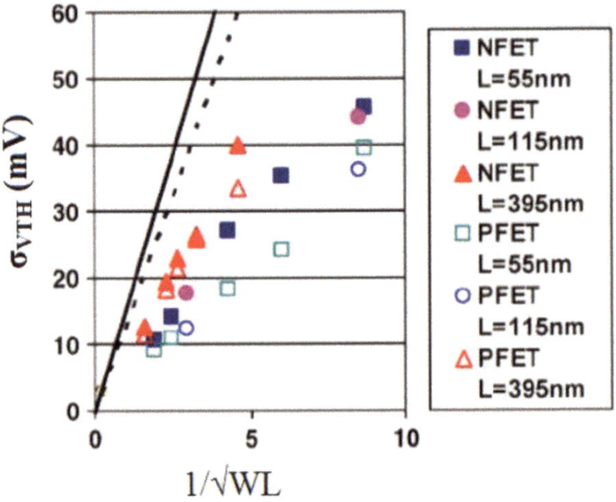

Fig. 7.3 Plot of σ_{VTH} in 65 nm technology as a function of $1/\sqrt{WL}$ [4]

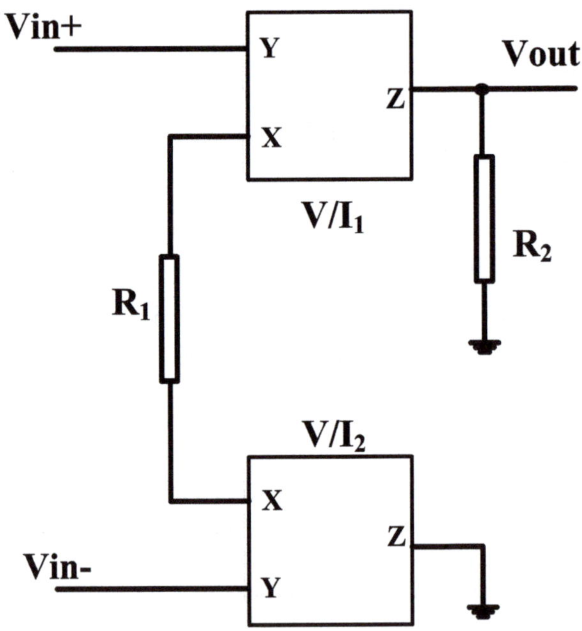

Fig. 7.4 Basic concept of voltage input CMIAs [5]

Inserting Eq. (7.6) into Eq. (7.5), the worst-case value of CMRR is:

$$CMRR = \frac{2-\varepsilon_1-\varepsilon_2}{2(\varepsilon_1+\varepsilon_2)} \tag{7.7}$$

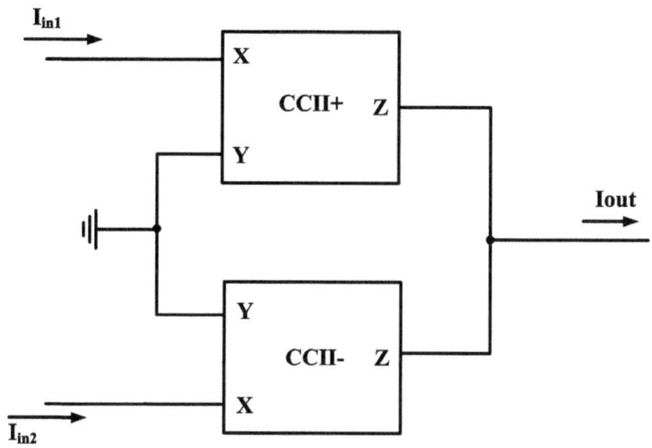

Fig. 7.5 A CCII based current input CMIA [6]

For example, for $\varepsilon_1 = \varepsilon_2 = 3\%$, 5% and 10%, the resulted CMRR is calculated as 24.17 dB, 19.55 dB and 13.89 dB, respectively. These results reveal that the value of CMRR in CMIAs is highly susceptible on matching between active building blocks.

Figure 7.5 shows a CCII based current input CMIA. Here, two different types of active building blocks, a CCII$^+$ and a CCII− are required to perform current subtraction. Providing matching between CCII$^+$ and CCII− is not as easy as the voltage input CMIAs where two similar types of active building block are used. In addition, in other structures involving resistor network such as [7–9], the matching between resistors is also important. Some CMIA topologies are based on single active element with differential input such those reported in [10–12]. In these structures, the difference between input signals is produced by active building block. For voltage input type, we have:

$$V_{out} = \alpha_1 V_1 - \alpha_2 V_2 \tag{7.8}$$

For current output type we have:

$$I_{out} = \beta_1 I_1 - \beta_2 I_2 \tag{7.9}$$

being α_i and β_i (for i = 1–2) internal parameters of the active building block which are close to unity.

In single active block structures, the accuracy of subtraction action performed by active building block determines the overall CMRR. Obviously, the matching between pair devices plays an important role in producing the difference between input signals.

7.3 Effective Techniques to Reduce Mismatch

Some layout techniques can be used to reduce the occurrence of mismatches. As all mismatches are inversely proportional to dimensions, large geometries will result in reduced mismatches. In this technique frequency performance is compromised. Several layout techniques such as common centroid and multiple fingers are used to improve matching. Other design rules such the use of end dummies or the minimum distance between matched pairs [13] are helpful in reducing mismatch between devices. Beside these techniques, the quality of fabrication process is also very essential to achieve a good matching between devices.

In supply current sensing CMIA designed for biomedical applications reported in [14], common centroid and interleaving techniques are applied to maintain good matching. Providing symmetry in the circuit is also very essential in mitigating mismatches. For example, in the supply current sensing CMIA, although the current of lower side is not used, the same current mirrors connected to the supply rails of Op-Amp_1 are connected to the supply rails of Op-Amp_2. This technique will improve the matching between the two input op-amps [14]. Here, a good matching is achieved at the expense of increased area.

Let us now consider current mirrors which are found in almost all current mode signal processing circuits. Any mismatch between the transistor parameter β (equal to $\mu C_{OX}W/L$) and V_{TH} causes inaccuracy in current copying operation. In the simple circuit of Fig. 7.6, the error between drain currents (I_{DS}) caused by mismatches between two transistors parameters is found in [3] as:

$$\left(\frac{\sigma(\Delta I_{DS})}{I_{DS}}\right)^2 = \left(\frac{\sigma(\Delta\beta)}{\beta}\right)^2 + 1/2\left(\frac{1}{V_{GS}-V_{TH}}\right)^2 \sigma^2(\Delta V_{TH}) \qquad (7.10)$$

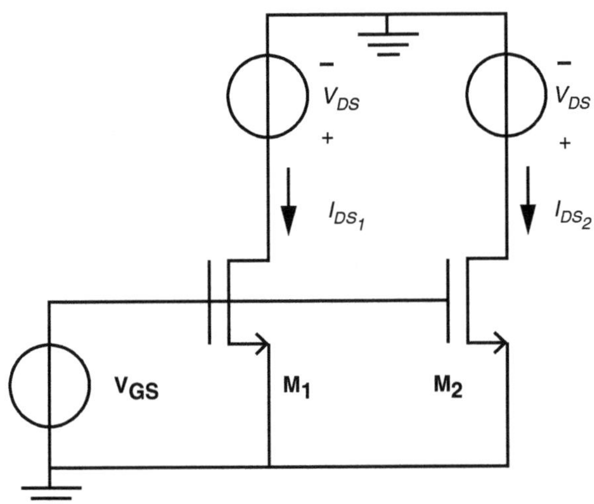

Fig. 7.6 Simple current mirror [3]

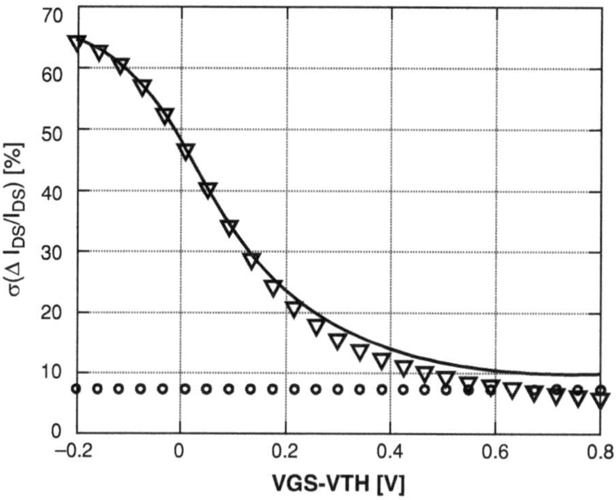

Fig. 7.7 Drain source current mismatch (solid curve) [3]

The plot of Eq. (7.10) for different bias points is shown in Fig. 7.7 in which the contributions of V_{TH} (∇) and β (°) are shown separately. For small values of $V_{GS} - V_{TH}$, the contribution of mismatch in V_{TH} is dominant over $\Delta\beta$. Increasing bias point reduces the effect of ΔV_{TH}. These results show a tradeoff between power consumption and accuracy.

7.4 A Current Input CMIA Topology with Robust Performance Against Mismatches

In [5], a current input CMIA topology is reported where the matching between two active building blocks has not a significant effect on the CMRR. The circuit is based on a Single Input Multiple Output Current Operational Amplifier (SI-MO COA), shown in Fig. 7.8. This COA is characterized by low input impedance (ideally zero), high output impedance (ideally infinite) and high current gain (ideally infinite). Due to the high current gain between input and output, it is possible to use COA in negative feedback configuration.

Figure 7.9 shows the COA-based CMIA topology. COA_2 is configured as current buffer. Its O_{1-} output is used to perform negative feedback. The relation between open loop gain (A_{02}) and closed loop gain (A_2) of COA_2 is:

$$A_2 = \frac{A_{02}}{1 + A_{02}} \tag{7.11}$$

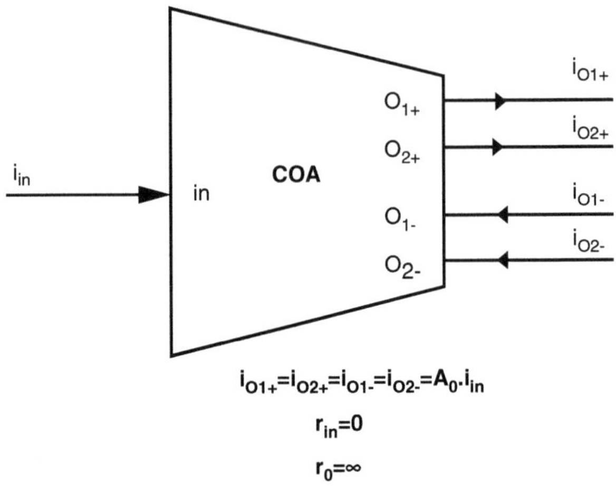

Fig. 7.8 Symbolic representation of SI-MO COA [5]

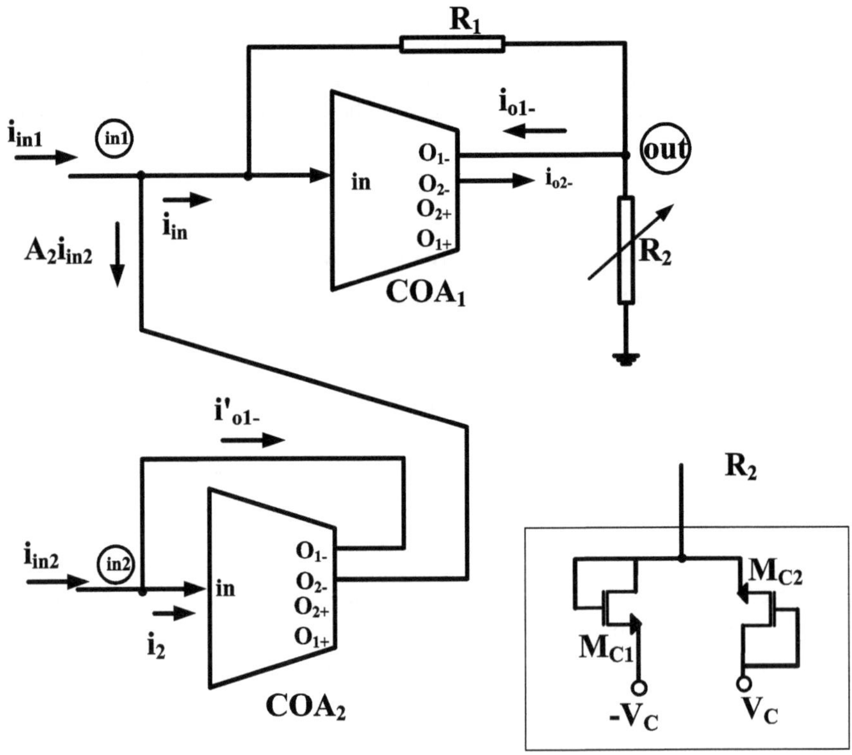

Fig. 7.9 Current input CMIA topology with robust performance against mismatches [5]

Therefore, for A_{02} >>1, the I_{in2} is transferred to O_{1-} and O_{2-} outputs by a gain close to unity. To subtract the O_{2-} output of COA_2 from I_{in1}, O_{2-} is directly connected to the input of COA_1. Therefore, the resulted input current to COA_1 is:

$$I_{in} = I_{in1} - A_2 I_{in2} \tag{7.12}$$

Using Eqs. (7.11 and 7.12), the common-mode input current (for $I_{in1} = I_{in2} = I_{cm}$), named I_{inc}, is given by:

$$I_{inc} \cong \frac{1}{A_{02}} I_{cm} \tag{7.13}$$

From Eq. (7.13) it is evident that common-mode inputs are attenuated before entering COA_1.

For differential-mode inputs ($I_{in1} = -I_{in2} = I_{dm}/2$), the input current i_{in} is named I_{ind} stated as:

$$I_{ind} = I_{dm} \tag{7.14}$$

COA_1 is configured as a current amplifier with gain of:

$$A_1 = \frac{i_{o2-}}{I_{in}} = \left(1 + \frac{R_1}{R_2}\right) \tag{7.15}$$

Both I_{ind} and I_{inc} are amplified by COA_1, so the overall CMRR is determined by the quality of current subtraction mechanism in input node. The resulted CMRR is expressed as [5]:

$$CMRR = A_{02} \tag{7.16}$$

Equation (7.16) states that CMRR is only determined by open loop gain of COA_2. Table 7.1 shows the simulated CMRR variation in 0.18 μm CMOS technology for different values of open loop gains of COA_1 and COA_2. For large variations in open loop gains, the CMRR variation is lower than 1 dB.

Table 7.1 CMRR variation with respect to A_{02} mismatches [5]

Open loop gains of		
COA_1 (A_{01})	COA_2 (A_{02})	CMRR
370	346	50.7 dB
370	383	51.5 dB

References

1. Drennan P. G., McAndrew C. C. (2003) Understanding MOSFET mismatch for analog design. IEEE Journal of Solid State Circuits, 38(3):450–456.
2. Agarwal K., Nassif S., Liu F., Hayes J., Nowka K. (2007) Rapid characterization of threshold voltage fluctuation in MOS devices. IEEE International Conference on Microelectronic Test Structures, Tokyo, 2007.
3. Kinget P. R. (2005) Device mismatch and tradeoffs in the design of analog circuits. IEEE Journal of Solid-State Circuits, 40(6):1212–1224.
4. Johnson J. B., Hook T. B., Lee Y. M. (2008) Analysis and modeling of threshold voltage mismatch for CMOS at 65 nm and beyond. IEEE Electron Device Letters, 29(7):802–804.
5. Safari L., Minaei S., Ferri G., Stornelli V. (2018) Analysis and design of a new COA-based current-mode instrumentation amplifier with robust performance against mismatches. AEU - International Journal of Electronics and Communications, 89:105–109.
6. Azhari S. J., Kaabi H. (2000) AZKA cell, the current-mode alternative of Wheatstone bridge. IEEE Transactions on Circuits and Systems I: Fundamental Theory and Applications, 47(9):1277–1284.
7. Pandey N., Nand D., Pandey R. (2016) Generalized operational floating current conveyor based instrumentation amplifier. IET Circuits Devices Syst., 10(3):209–219.
8. Pandey N., Nanand D., Venkatesh Kumar V., Kumar Ahalawat V., Malhotra C. (2016) Realization of OFCC based transimpedance mode instrumentation amplifier. Theoretical and Applied Electrical Engineering, 14:162–167.
9. Pandey R., Pandey N., Paul S. K. (2013) Electronically tunable transimpedance instrumentation amplifier based on OTRA. Journal of Engineering, 10:, Article ID 648540, 5 pages.
10. Gupta K., Gupta P., Pandey N., Pandey R. (2016) CDBA - current based instrumentation amplifier. Journal of Communications Technology, Electronics and Computer Science, 4:11–15.
11. Cini U. (2014) A low-offset high CMRR current-mode instrumentation amplifier using differential difference current conveyor. 1st IEEE International Conference on Electronics, Circuits and Systems (ICECS), Marseille, 2014.
12. Hassan T. , Mahmoud S. A. (2010) New CMOS DVCC realization and applications to instrumentation amplifier and active-RC filters. International Journal of Electronics and Communcations (AEÜ), 64:47–55, 2010.
13. Laker K. R., Sansen W. M. C. (1994) Design of analog integrated circuits and systems. Spinger, 1994.
14. Douglas E. L., Lovely D. F. , Luke D. M. (2004) A low-voltage current-mode instrumentation amplifier designed in a 0.18-micron CMOS technology.Canadian Conference on Electrical and Computer Engineering 2004 (IEEE Cat. No.04CH37513), 2004.

Chapter 8
CMIA for Biomedical and Low-Voltage Low-Power Applications

8.1 CMIA for Biomedical Applications

8.1.1 CMIA Design Considerations for Biomedical Data Acquisition Systems

The bio-potential signals produced in the human body provide some useful information about the health condition of various organs. An accurate and robust acquisition of these signals is necessary in healthcare systems. For many years, IAs are used for this aim. For implantable biomedical applications, CMIA is more appropriate because it does not require resistor matching to achieve high CMRR. In addition, current-mode circuits are featured by low voltage low power operation making CMIA an appropriate choice for battery powered and portable biomedical applications.

Bio-potential signals such as electrocardiogram (*ECG*) and electroencephalogram (*EEG*) have low values in the range of a few μV to a few mV and very low frequencies [1]. To measure these low value signals in the noisy environment, high CMRR and PSRR are demanded [1]. In addition, due to the very low value of bio potential signals, the IA for these applications must exhibit very low input referred noise. As the operating frequency is low, flicker noise is the dominant noise source. In this sense, even if the PMOS transistors produce low flicker noise [2], because they show larger threshold voltage compared to NMOS transistors, they are not preferred in low-voltage applications. Another problem is the DC component caused by electrodes which is in the range of several mV. This DC component limits the dynamic range which is an important aspect in low voltage circuits. To remove the DC component, a high pass filter with low cut-off frequency is required at the IA input stage. In addition, to mitigate the attenuation due to electrode impedance and mismatch a very high input impedance is required [1, 3, 4]. Due to continuously decreasing voltage supplies, high voltage swings at output and input stages are also required.

G. Ferri et al., *Current-Mode Instrumentation Amplifiers*, Analog Circuits and Signal Processing, https://doi.org/10.1007/978-3-030-01343-1_8

8.1.2 CMIA Based on Bootstrapping Technique for ECG and EEG Applications

A standard biomedical recording system utilizes three electrodes, two of which are attached on the patient body operating as sensing electrodes, while the third electrode is used to perform a ground connection for safety reasons. This electrode is connected to a Common Mode Feed-Back (*CMFB*) system, to reduce the effect of common-mode noises and disturbances. Especially for fetal monitoring where inspection must be less invasive as possible, it is preferable to use two electrode system and remove the CMFB one [3, 4]. As the high input impedance is necessary to reduce the effect of mismatch between two sensing electrodes, if the third electrode is removed, a suitable technique should be applied to provide high input impedance.

Bootstrap is a well-known technique in analog circuits to increase the input impedance by using a small amount of positive feedback. In [3, 4], this technique is used to design a two-electrode current-mode biomedical recording amplifier which can provide both high input impedance and ground connection without the need for extra third electrode.

Figure 8.1 shows the two electrodes bootstrapped CMIA based on four CCIIs [3].

Parasitic resistances at X (R_x) and Z (R_z) nodes are shown. In this circuit, the ground connection is ensured by resistors R_b and R_{GND} in lower side and by R'_b and R'_{GND} in the upper side. The voltage buffer action between Y and X nodes of $CCII_1$-$CCII_2$ in the upper side of the circuit ideally gives zero voltage drop across R'_b.

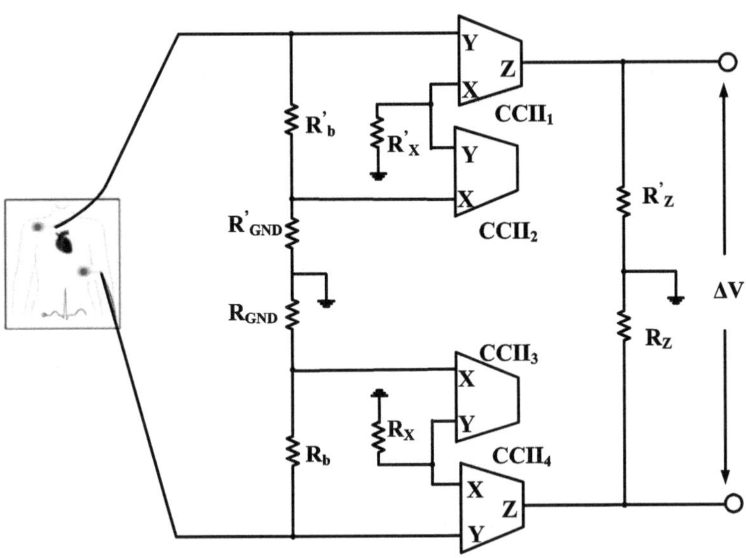

Fig. 8.1 Four CCII based bootstrapped CMIA [4]

Therefore the current flowing through R'_b-R'_{GND} is ideally zero. The same explanation holds for lower side. Zero current following through R_b-R_{GND} and R'_b-R'_{GND} means the system input impedance Z_{in} has ideally an infinite value. Considering the CCIIs non-idealities, the input impedance is [3]:

$$Z_{in} = \frac{R'_b}{1-\alpha_1\alpha_2} + \frac{R_b}{1-\alpha_3\alpha_4} \tag{8.1}$$

where α_i (for i = 1–4) is the voltage gain between Y and X terminals of corresponding CCII. In [3], using the 0.35 μm CMOS technology, the value of about 45 MΩ is achieved in the frequency range of 0–10 Hz.

The voltage gain is given by:

$$A_v = 2\beta \frac{R_z}{R_x} \tag{8.2}$$

The improved version of the bootstrap circuit of Fig. 8.1 is reported in [4]. The circuit is shown in Fig. 8.2. In order to further increase the input impedance and decrease the power consumption, the number of CCIIs is reduced to two. In this circuit ground connection is provided by R'_b–R'_{GND} in the upper side and R_b–R_{GND} in the lower side. The voltage buffer action between Y and X terminals of $CCII_1$ ensures ideally zero voltage drop on R'_b. Similarly, the voltage drop on R_b is kept at zero by $CCII_2$ in the lower side. The input impedance is:

$$Z_{in} = \frac{R'_b}{1-\alpha_1} + \frac{R_b}{1-\alpha_2} \tag{8.3}$$

Comparing input impedance of four and two CCII based bootstrapped CMIAs (Eqs. 8.1 and 8.3) reveals that the latter one provides higher input impedance. Using

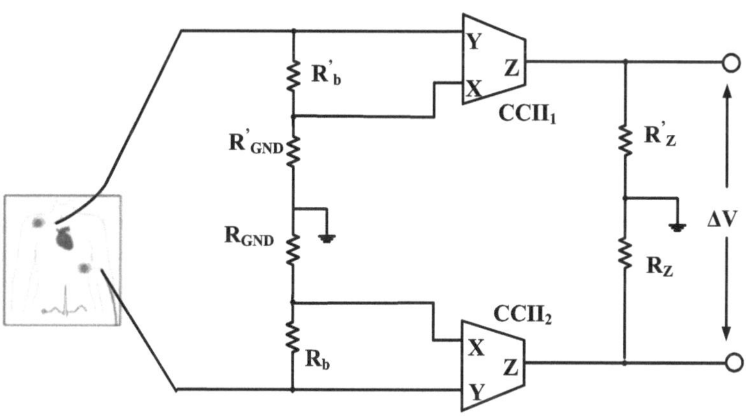

Fig. 8.2 Two-CCII based bootstrapped CMIA [4]

the CCII reported in [5] and 0.35 μm CMOS technology, for the two CCII based circuit, input impedance of 62 MΩ is achieved at 0–50 Hz. The gain of the two CCII bootstrapped CMIA is also:

$$A_v = 2\beta \frac{R_z}{R_{GND}} \tag{8.4}$$

In [6], the design guidelines on noise analysis and optimization of the four CCII bootstrapped CMIA of Fig. 8.1 are reported.

8.1.3 CMIA with Very Low Input Noise Voltage

A low-voltage low-noise IA suitable for portable biomedical applications is designed in [7]. The main features of this circuit are its capability to work under a single 1-V supply and has low input referred noise. Figure 8.3 shows its operation principle. Although this circuit is not based on current-mode active building blocks, its operation is based on current-mode signal processing. This fact allows us to categorize it in CMIA group. The noise reduction techniques employed in [7] are very helpful in the design of CMIAs. As Fig. 8.3 shows, the input voltages are converted to a proportional current by R_i. This current is amplified by a gain of k and then at the output it is converted again to voltage by R_o resulting in a total gain of:

$$A_V = k \frac{R_o}{R_i} \tag{8.5}$$

Transistor level implementation of Fig. 8.3 is shown in Fig. 8.4. Transistors M_1-M_{12} constitute an operational transconductance amplifier where M_1-M_{10} form a series-shunt feedback loop [7]. M_1 and M_2 operate as voltage followers and copy input voltage onto R_i and convert it into a current equal to:

$$i_{Ri} = \frac{v_{in}}{R_i} \tag{8.6}$$

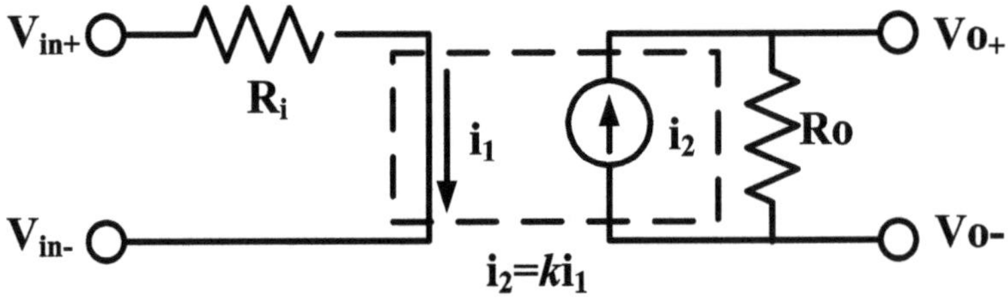

Fig. 8.3 The operation principle of the CMIA in [7]

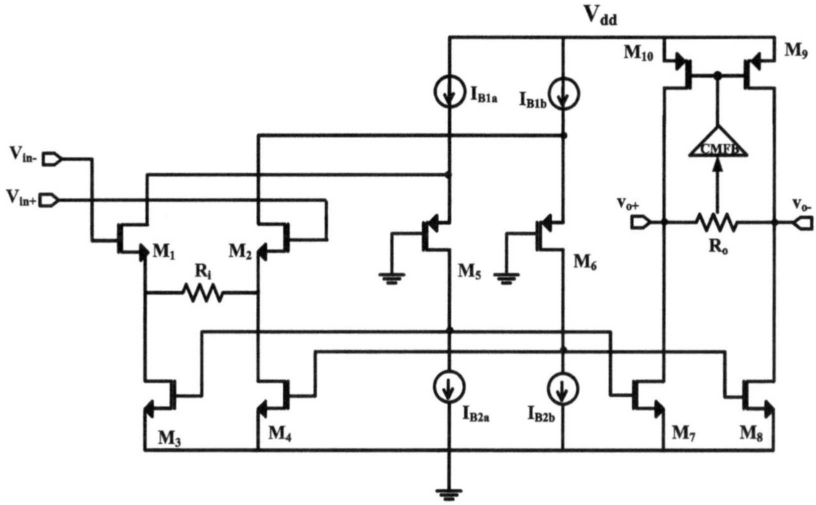

Fig. 8.4 Transistor level implementation of the CMIA reported in [7]

Transistors M_3-M_4 are identical to M_7-M_8. Therefore, current i_{Ri} is transferred to R_o resulting in a voltage gain of:

$$A_v = \frac{R_o}{R_i} \tag{8.7}$$

The input referred thermal and flicker noises of this circuit are derived in [7] as:

$$\overline{V^2_{n,thermal}} \approx 4kT\Delta f \left[\begin{array}{l} \frac{1}{3}\left(\frac{2}{g_{m1}} + R_i\right)^2 \cdot \left(g_{m(Ib1,A)} + g_{m(Ib2,A)}\right) \\ + \frac{4}{3g_{m1}} + \frac{2g_{m3} + g_{m9}}{3} . R_i^2 + R_i + \frac{R_i^2}{R_{out}} \end{array} \right] \tag{8.8}$$

$$\begin{aligned} \overline{V^2_{n,flicker}} \approx & \frac{1}{2}\left(\frac{2}{g_{m1}} + R_i\right)^2 \cdot \left(g^2_{m(Ib1,A)} . \overline{V^2_{nf(Ib1,A)}} + g^2_{m(Ib1,A)} . \overline{V^2_{nf(Ib1,A)}}\right) \\ & + 2\overline{V^2_{nf1}} + \left(g^2_{m3} . \overline{V^2_{nf3}} + \frac{1}{2} g^2_{m10} \overline{V^2_{nf3}}\right) . R_i^2 \end{aligned} \tag{8.9}$$

where $\overline{V^2_{nfi}}$ denotes the flicker noise of M_i transistor and is defined as:

$$\overline{V^2_{nfi}} = \frac{K_i}{(W.L)_i f} \Delta f \tag{8.10}$$

In the CMIA in [7], the following points are considered to achieve low input noise under low supply voltage:

1. The value of g_{m1} (transconductance of transistor M_1-M_2) is chosen as larger as possible while those of M_3, M_4, I_{B1a}, I_{B1b} and M_7, M_8, I_{B2a}, I_{B2b} are set as lower as possible, while the related aspect ratios are determined by the available headroom.
2. As the transistors M_5-M_6 are in cascode configuration with M_1-M_2, their contribution in input noise is negligible therefore, the aspect ratios of these transistors can be chosen large to reduce their overdrive voltage and improve voltage headroom.
3. To reduce the noise of M_3, M_4, I_{B1a}, I_{B1b} and M_7, M_8, I_{B2a}, I_{B2b}, they should operate in saturation region. Therefore, the chosen sizes must ensure all these transistors to remain in saturation region.

8.2 Low-Voltage Low-Power CMIAs

8.2.1 CMIA with Rail-to-Rail Input/Output Common-Mode Range

8.2.1.1 CCII Based CMIA with Rail-to-Rail Input

An input rail-to-rail stage is designed in [8] based on the Wilson CMIA [9] topology (Fig. 8.5). Here, due to the voltage buffer action between Y and X terminals of $CCII_1$ and $CCII_2$, the input voltage is transferred across R_1 and is converted to current. The produced current is transferred to Z terminal and converted to voltage at output

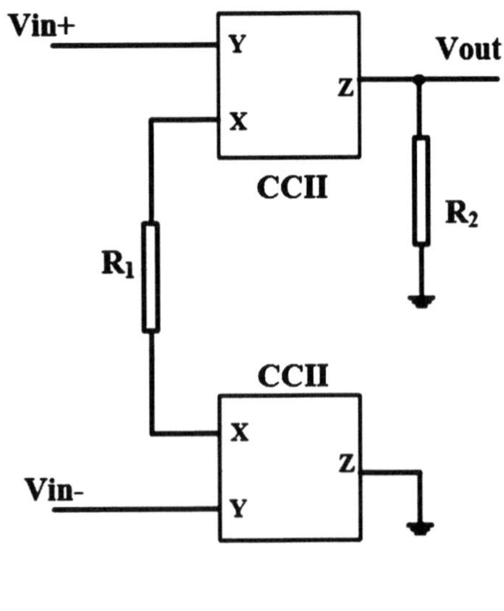

Fig. 8.5 The Wilson CMIA structure [9]

node. By neglecting parasitic impedances at X terminals of CCIIs the output voltage is:

$$v_{out} = \frac{\beta_1 \left(\alpha_1 v_{in+} - \alpha_2 v_{in-}\right) R_2}{R_1} \tag{8.11}$$

where α_1 and α_2 are the voltage tracking gain between Y and X terminals of $CCII_1$ and $CCII_2$ respectively and β_1 is the current tracking gain between X and Z terminals of $CCII_1$.

In this CMIA, rail-to-rail operation requires that the used CCIIs have the complete common-mode input range. The implementation of rail-to-rail CCIIs is shown in Fig. 8.6. The input stage consists of two complementary source coupled pairs connected in parallel. The output signals of the input pairs are connected to the folded cascode stage consisting of transistors M_{12}-M_{19} that produces a single-ended differential output. Transistors M_1-M_4 are biased in weak inversion resulting in a total transconductance, g_{mT}, proportional to bias current and equal to [8]:

$$g_{mT} = \frac{I_p}{2n_p V_T} + \frac{I_n}{2n_n V_T} \tag{8.12}$$

being I_p and I_n the bias currents in P-channel and N-channel pairs, respectively. By assuming $n_n \approx n_p$, g_{mT} can be made constant if the sum of I_p and I_n is kept constant. This process is performed by transistor M_7 which operates as a simple switch. For low common-mode input voltages, M_7 is off, therefore I_{ref} flows into P-type pair (M_3-M_4). As common-mode voltage increases beyond V_B, M_7 is switched on, and I_{ref} is injected to N-type pair (M_1-M_2). In any case, the sum of bias currents of P-type and N-type pairs are constant at I_{ref} resulting in a constant g_{mT} in whole range of common-mode input voltage range. The negative feedback loop established by transistors M_{20}-M_{21} which operate as inverter, provides low input impedance at X

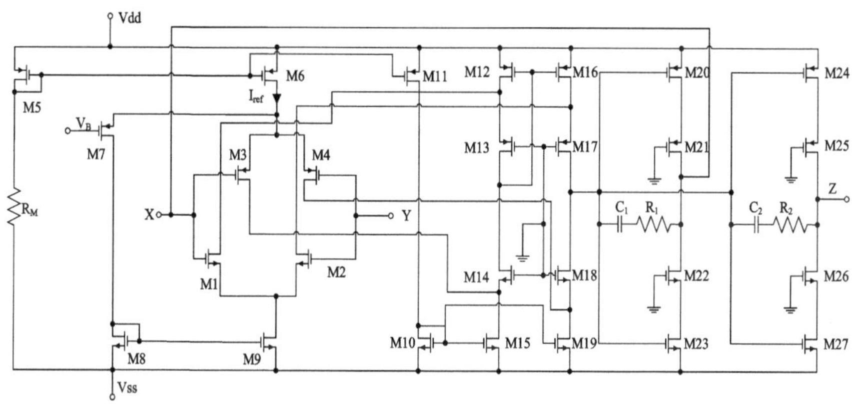

Fig. 8.6 Constant g_m rail-to-rail input CCII [8]

terminal and close to unity voltage gain between Y and X terminals. The second inverter, made of M_{24}-M_{27}, copies the X terminal current to Z terminal and provides high impedance at Z terminal. The CMIA of [9] based on rail-to-rail CCII of [8] has low frequency CMRR of 105 dB, power consumption of 177 μW in 0.35 μm CMOS technology for a single supply voltage of 1.5 V.

8.2.1.2 CMIA Based on Supply Current Sensing Technique with Rail-to-Rail Input and Output

In [10] a CMIA based on supply current sensing topology is designed. It exhibits improved input and output common-mode ranges. The CMIA topology is shown in Fig. 8.7. In this circuit two solutions are adopted to achieve near rail-to-rail operation. First, the required overdrive voltage of the current mirrors which are connected in series with Op-Amp supply rails to transfer the output current of Op-Amp to the output stage of CMIA are avoided. This is done by using trans-conductance amplifiers made of transistors M_{15}-M_{16} and M_{17}-M_{18} instead of current mirrors. Then, the used Op-Amps are designed to have rail-to-rail operation.

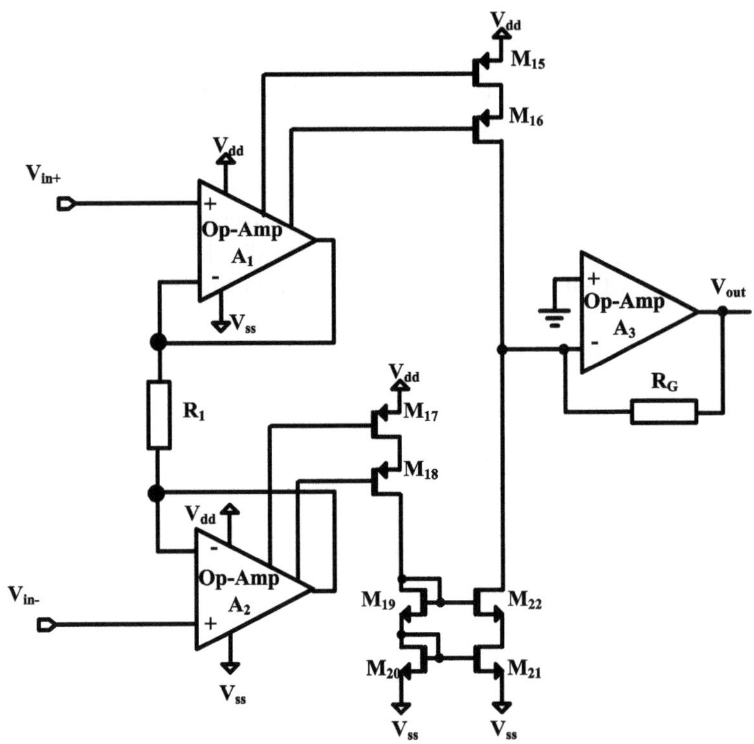

Fig. 8.7 CMIA structure reported in [10]

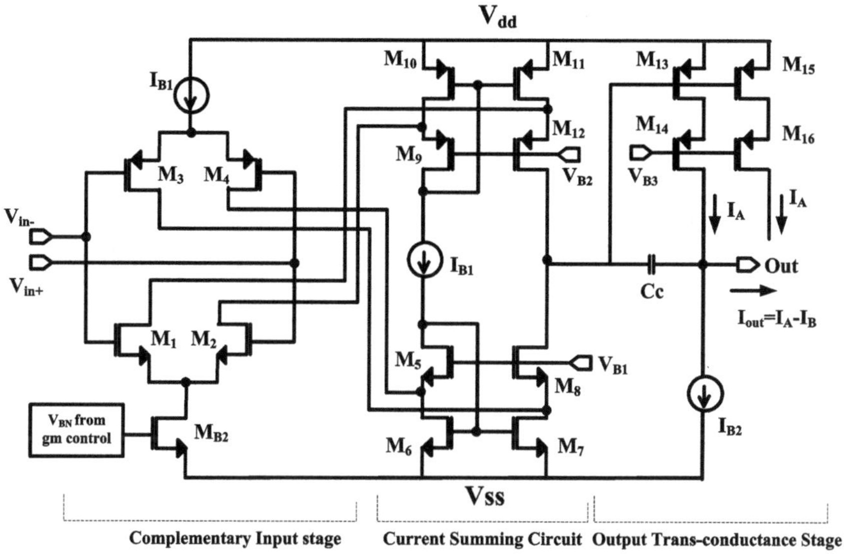

Fig. 8.8 The operational amplifier structure used in [10]

The Op-Amp internal topology is shown in Fig. 8.8 [10, 11]. To have rail-to-rail common-mode input voltage range, complementary differential pairs are used. To ensure good harmonic and stability performance the sum of trans-conductances of both pairs are kept constant at g_{mREF} [10]:

$$g_{mREF} = g_{mP} + g_{mN} \tag{8.13}$$

where terms g_{mP} and g_{mN} are the transconductances of P-type and N-type differential pairs [10]. To satisfy Eq. (8.13), the values of g_{mP} and g_{mN} should vary in opposite directions in order to keep their sum equal to constant value of g_{mREF}. For this purpose, the bias current of N-pair is controlled by a g_m control circuitry. Figure 8.9 shows the circuit used to keep the g_m of complementary pair at a constant value. The P-type and N-type differential pairs in control circuitry are identical to those employed in the Op-Amps input stage. The inputs of differential pairs P_1, P_{REF} and N_1 are biased in linear region by connecting their inputs to DC voltage V. The inputs of P_0 are connected to input voltages therefore its bias current varies by input common-mode voltage. The bias current of P_1 is also dependent on the input common-mode voltage because it is biased by a replica of P_0 bias current. The inputs of P_{REF} pair are connected to a constant voltage making its bias current independent of input common-mode voltage. The output currents of P_1 and P_{REF} are added to produce the voltage (V_{BN}), which controls the bias current of N pair so to keep the total g_m of complementary pair as constant as possible.

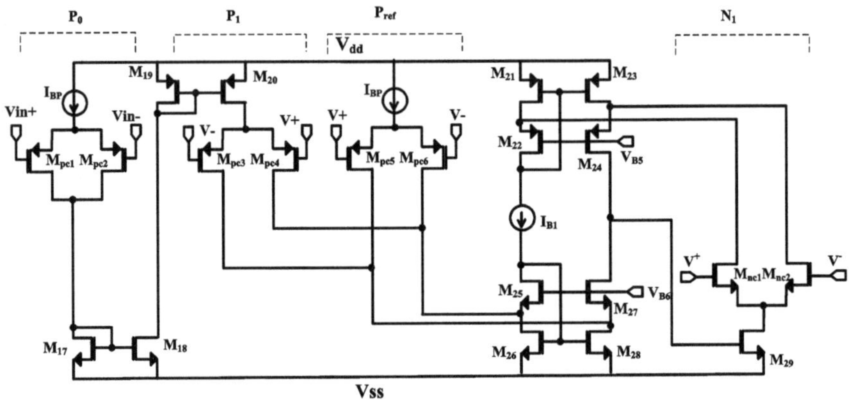

Fig. 8.9 g_m control circuit used in [10]

Using the 0.6 μm XC06 XFAB CMOS technology and supply voltage of ±1.5 V, a value of 61.9 dB CMRR is measured for the CMIA of Fig. 8.7. The input common-mode range is 86% supply rails and the output is full. To increase the value of CMRR, the input Op-Amps (A and B) are required to have a high CMRR value [10]. In another design, the CMRR value of input Op-Amps is improved and as a result, the CMRR of CMIA is increased to 93 dB. This is done by relaxing the design constraints on Op-Amp C and replacing it with a simple low-power two stage Op-Amp because rail-to-rail input is not required for it. In addition, a single control circuitry is adopted for both Op-Amp A and B. With these two changes, more power consumption can be dedicated to input Op-Amps while the total power consumption is kept in the desired range. The CMRR of Op-Amp A and B is improved by adopting appropriate values for bias current and aspect ratios of their input transistors resulting in a higher CMRR for the whole CMIA.

8.2.1.3 CMIA with Rail-to-Rail Input Voltage and Very Low Offset Based on Transmission Gate Chopper Switching

In [12], by combining current-mode and chopper modulation techniques a CMIA with rail-to-rail input voltage range and very low input offset is designed. The circuit employs DVCC as active building block (Fig. 8.10), transmission gate based chopper switching and AC coupled interfacing at the input of amplifier. Compared to other chopper modulated IAs using voltage-mode technique such as [13–15], employing DVCC as core amplifier in the CMIA of [12] reduces the circuit complexity, thanks to a simpler implementation of algebraic functions in current domain.

Figure 8.11 shows the general schematic of the CMIA of [12]. The input signal is chopper modulated, then the AC part of the modulated signal is fed into main

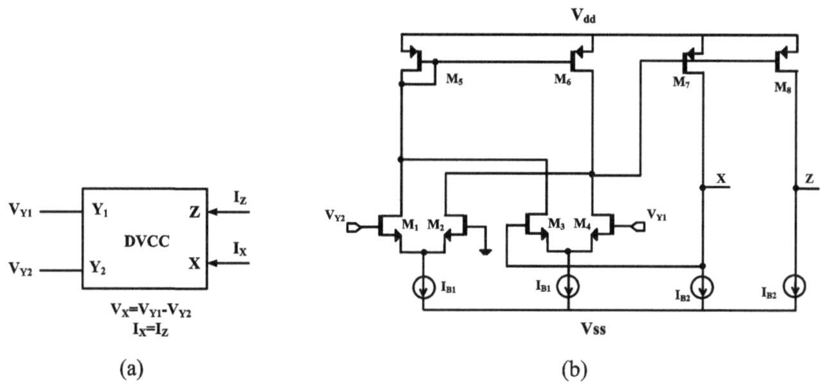

Fig. 8.10 DVCC (a) Symbol (b) Internal circuit [12]

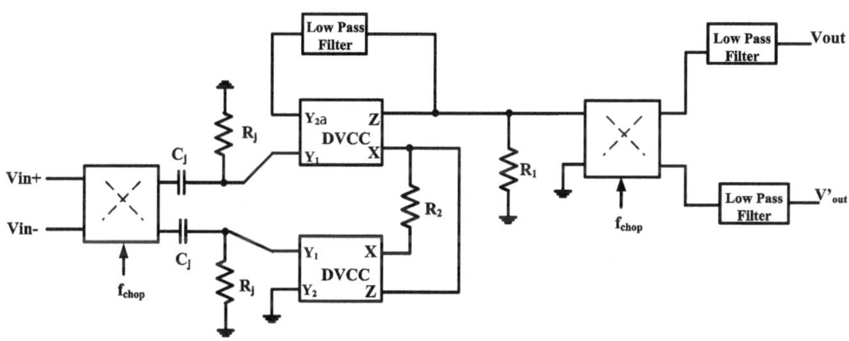

Fig. 8.11 Rail-to-rail, low offset hopper modulated CMIA implementation [12]

amplifier input using capacitive coupling circuit where it is amplified by the amplifier stage. The output of amplifier is low-pass filtered by the residual offset removal path to produce the offset voltage of the modulated signal which is fed back to the amplifier input and subtracted. This process totally removes the offset and noise voltages at the amplifier output. The produced clean output signal of amplifier is demodulated and low pass filtered at the output. A complementary output can also be produced as is shown in Fig. 8.11. To have rail-to-rail input common-mode range, complementary transmission gates shown in Fig. 8.12 employ the used chopper switches. The simulated value of CMRR for this circuit is 107 dB and the input referred offset voltage is only 0.9 μV. Without the offset removal path, the offset increases to 400 μV.

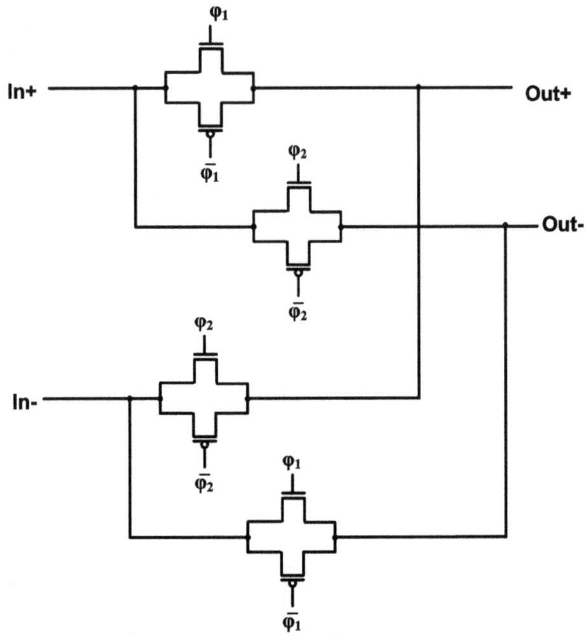

Fig. 8.12 Chopper modulation using complementary transmission gates [12]

8.2.2 Low Voltage Implementation of CMIAs Based on Supply Current Sense Technique

8.2.2.1 Supply Current Sensing CMIA Using FVF Based Double Current Sense Technique

In [16] using a FVF based double current sense technique, the required supply voltage for CMIA using supply current sense topology is reduced. The general schematic is shown in Fig. 8.13. In the conventional circuit, the gate of M_2 is connected to a constant bias voltage, therefore i_x is provided by one transistor. However in [16], the gate connection of M_2 is changed as is shown in Fig. 8.14 that shows the internal structure of Op-Amp. By this modification, i_x is provided by both M_1 and M_2. Both the sourced and sank current are sensed by FVF based current mirrors. The drain of M_1 is connected to the input of FVF based P-Type current mirror made of M_{p1}-M_{p3}. By this connection M_1 can sense the sourced current of the voltage follower inside Op-Amp. The source of M_2 is connected to input of FVF based N-Type current mirror made of M_{n1}-M_{n3} and can sense the sank current of voltage follower inside Op-Amp. Both the sourced and sank currents are added at the output of FVF based current mirrors. The i_x sourced from V_{DD} and sank into V_{SS} is transferred to the input of Op_3 by FVF based current mirrors. With double current sense technique, the GBW of CMIA is also improved because the input impedance of FVF current mirror has larger cutoff frequency when compared to other conventional current mirrors [16]. Using a 0.5 μm On-Semiconductor CMOS technology and supply voltage

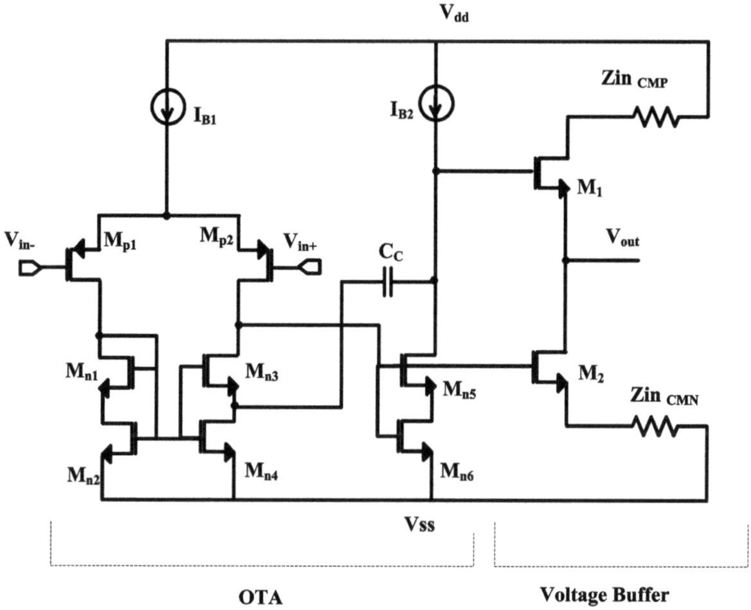

Fig. 8.14 Internal structure of the used Op-Amps in [16]

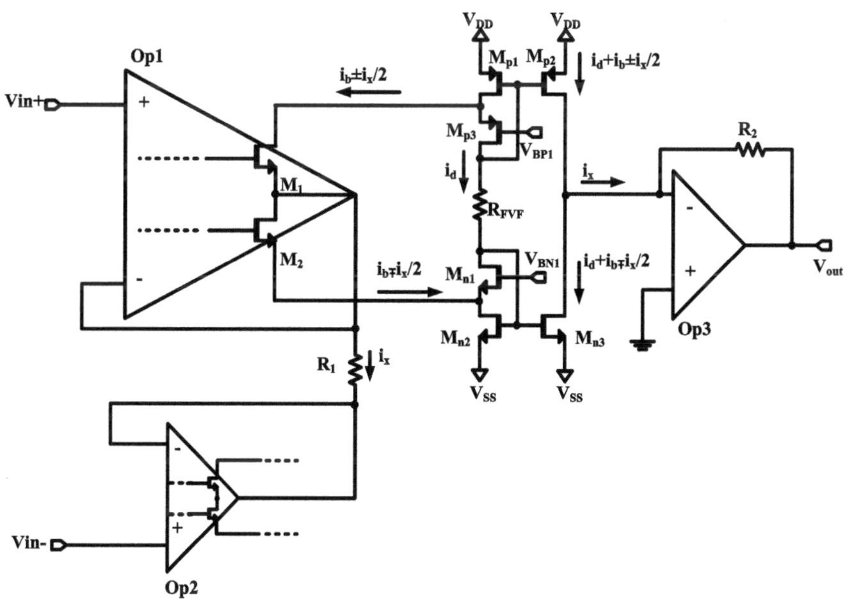

Fig. 8.13 CMIA with FVF based double current sense technique [16]

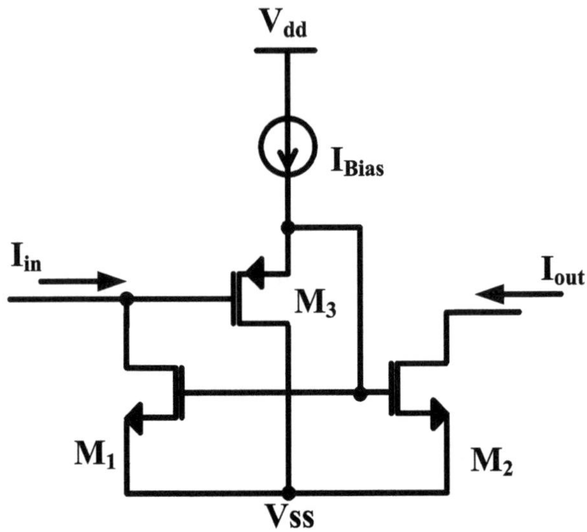

Fig. 8.15 Low-voltage level shifted current mirror [17]

of ±1.65, a value of 96 dB is achieved for CMRR with power consumption of 313.5 μW.

8.2.2.2 Supply Current Sensing CMIA Using NFET Based Low Voltage Current Mirrors

In [17] two techniques are used to reduce the required supply voltage of CMIA based on supply current sensing technique. First, the low-voltage current mirror shown in Fig. 8.15 is used for current sensing. In this current mirror the level shifter implemented with a simple source follower reduces the current mirror required input voltage. Second, the "native" or "natural FETs" (NFET) are used at the Op-Amps output source follower stage. The NFET is a FET with a reduced threshold voltage that is obtained by an extra step of intentionally doping the channel region [17]. The NFETs in 0.18 μm CMOS technology used in [18] has a threshold voltage of 50 mV. At a supply voltage of 1.8 V, a CMRR of 143 dB has been obtained in simulations.

8.2.2.3 Supply Current Sensing CMIA Using FVFs as Input Voltage Buffer and Resistor Based Current Mirror

Figure 8.16 shows the simplified structure of the CMIA described in [18], which achieves low-voltage low-power operation, where simple voltage buffers are used. Instead of Op-Amps in unity gain configuration, simple FVF based and Super transistor-based voltage buffers shown in Fig. 8.17 are utilized. As the current mirror is

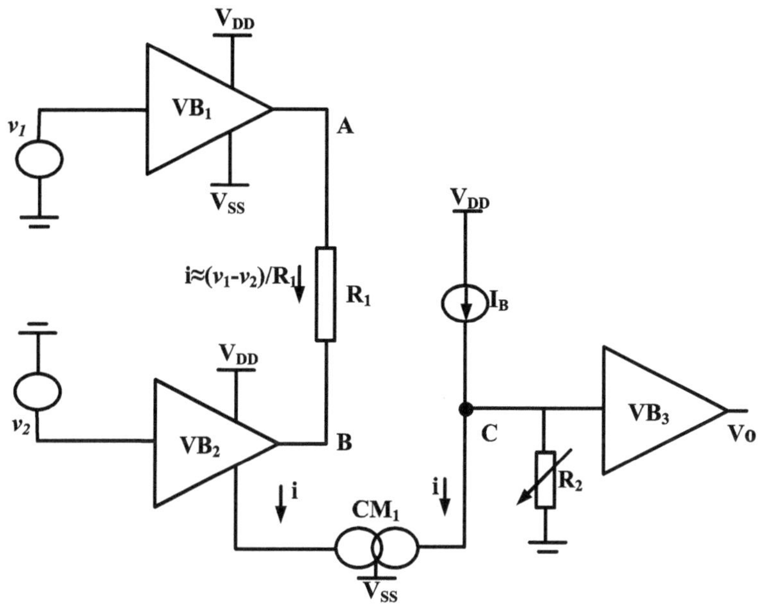

Fig. 8.16 simplified schematic of CMIA in [18]

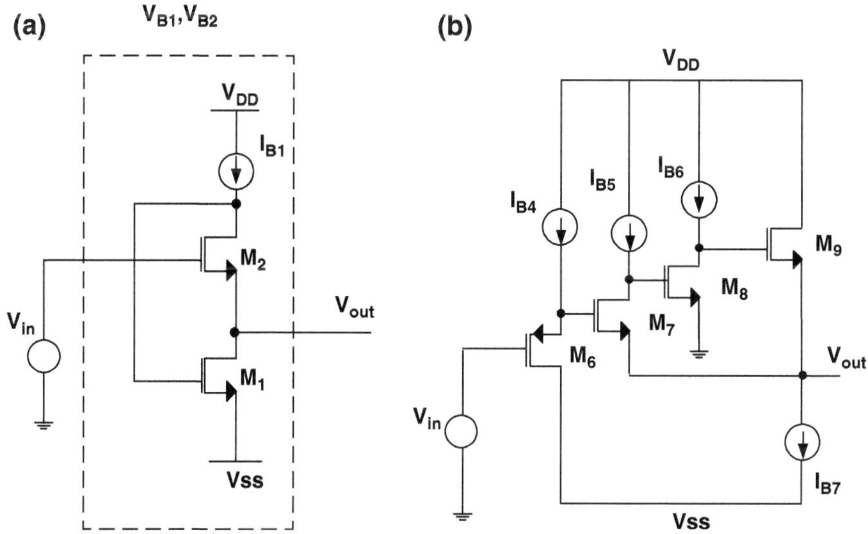

Fig. 8.17 The used (**a**) FVF based and (**b**) super transistor-based voltage buffers [18]

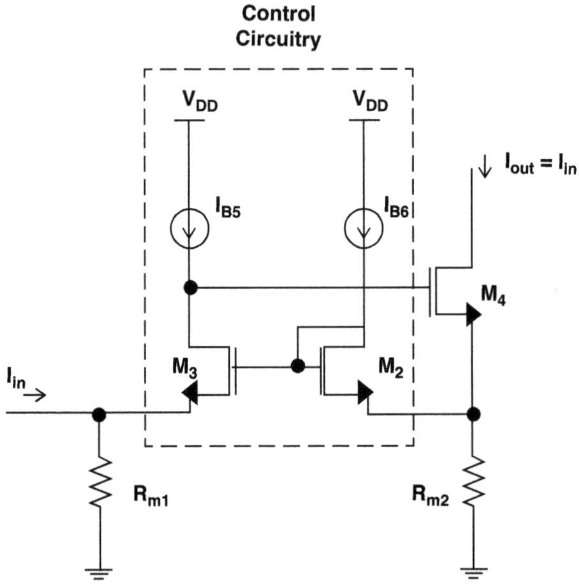

Fig. 8.18 Resistor-based current mirror [19]

connected in series with the supply rail of V_{B2}, to reduce the required supply voltage, the resistor-based current mirror shown in Fig. 8.18 [19] which has a very low input voltage requirement (in the range of a few mV) is used. Here, the output current of V_{B2} is injected into low value resistor R_{m1} and is copied to node C. The increase in supply voltage is equal to the voltage drop across R_{m1} which is negligible for low value resistor. The complete schematic of the low-voltage low-power CMIA of [18] is shown in Fig. 8.19. For this circuit, for a 0.18 μm CMOS technology and supply voltage of ±0.9 V, the value of 71 dB for CMRR and the power dissipation of 770 μW are reported.

8.2.3 *An OFCC-Based CMIA with 0.4 V Supply Voltage*

In [20], a very low-voltage low-power OFCC based CMIA is reported. The energy efficient subthreshold region is utilized in the design of OFCC circuit. In subthreshold region, the highest bandwidth is achieved when compared to other operating regions and for low supply voltages. Therefore, the subthreshold region is the highest energy efficient region for transistors. In addition, to bias the transistors of OFCCs in subthreshold region, self cascode technique is utilized to improve the current tracking feature. To reduce the flicker noise, the CMIA of [20] employs chopping technique. Figure 8.20 shows the schematic of the CMIA of [20] which is composed of two OFCCs, two feedback resistors named R_W, a gain setting resistor named R_G, load resistors R_L, input high pass filter (HPF), output low-pass filter (LPF) and chopper switches.

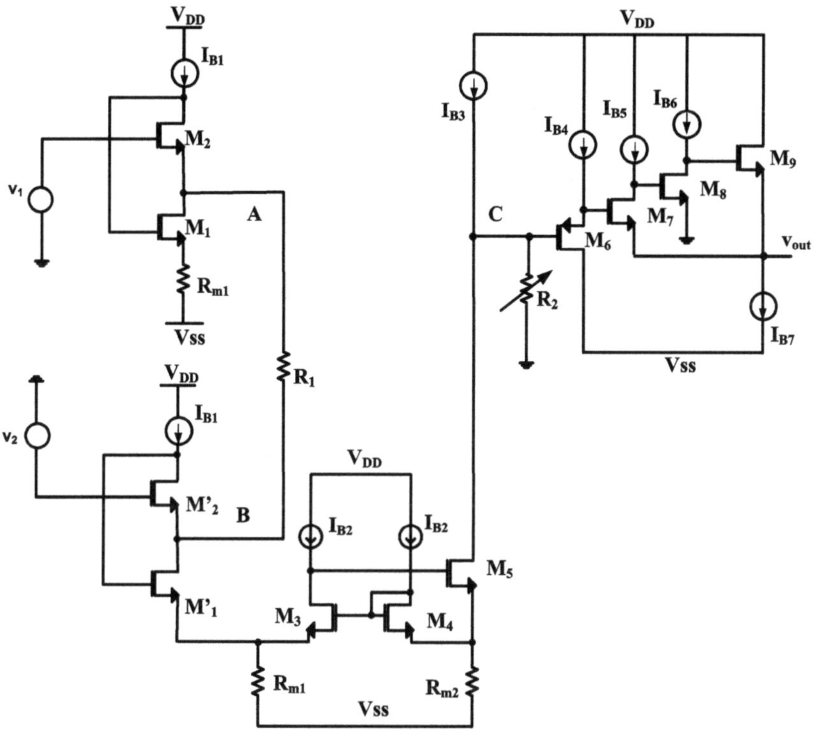

Fig. 8.19 Complete schematic of the CMIA in [18]

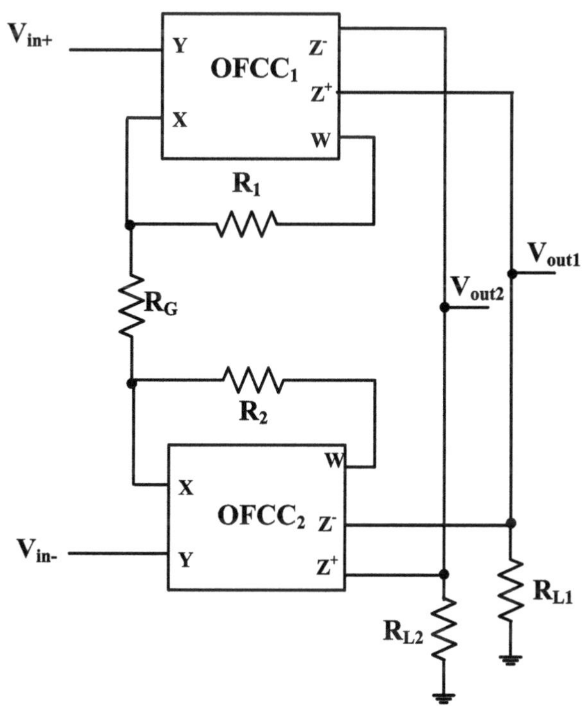

Fig. 8.20 The Low-Voltage Low-Power CMIA of [20] designed in the subthreshold region

Two versions of OFCCs are designed in 90 nm and 130 nm CMOS technologies and both circuits operate under low supply voltage of 0.4 V. The power consumptions are 11 μW and 14 μW for 90 nm and 130 nm, respectively. The simulated CMRR are reported as 76 dB and 64.7 dB for 90 nm and 130 nm technology, respectively.

References

1. Das D. M., Ananthapadmanabhan J., Baghini M. S., Sharma D. K. (2014) Design considerations for high-CMRR low-power current mode instrumentation amplifier for biomedical data acquisition systems. IEEE International Conference on Electronics, Circuits and Systems (ICECS).
2. Jakobson C., Bloom I., Nemirovsky Y. (1998) 1/f noise in CMOS transistors for analog applications from subthreshold to saturation. Solid-State Electronics, 42(10):1807–1817.
3. Ferri G., Stornelli V., Di Simone A. (2011) A CCII-based high impedance input stage for biomedical applications. Journal of Circuits, Systems, and Computers, 20(8):1441–1447.
4. Stornelli V., Ferri G. (2014) A single current conveyor-based low voltage low power bootstrap circuit for electroCardioGraphy and electroEncephaloGraphy acquisition systems. Analog Integrated Circuits and Signal Processing, 79(1):171–175.
5. Ferri G., Stornelli V., Fragnoli M. (2006) An integrated improved CCII topology for resistive sensor application. Analog Integrated Circuits and Signal Processing Journal, 48(3):247–250.
6. Kalogiros S., Noulis T. (2016) Noise analysis and optimization of CMOS CCII+ based ECG Systems. International Conference on Telecommunications and Signal Processing (TSP), Vienna.
7. Wu H., Xu Y. P. (2005) A low-voltage low-noise CMOS instrumentation amplifier for portable medical monitoring systems. International IEEE-NEWCAS Conference.
8. Stornelli V., Ferri G., Pantoli L., Barile G., Pennisi S. (2018) A rail-to-rail constant-gm CCII for instrumentation amplifier applications. AEU - International Journal of Electronics and Communications, 91:103–109.
9. Wilson B. (1989) Universal conveyor instrumentation amplifier. Electronics Letters, 25(7):470–471.
10. Vieira F. C., Prior C. A., Rodrigues C. R., Perin L., Martins J. B. (2008) Current mode instrumentation amplifier with rail-to-rail input. Analog Integrated Circuits and Signal Processing, 57: 29–37.
11. Prior C. A., Vieira F. C. B., Rodrigues C. R. (2006) Instrumentation amplifier using robust rail-to-rail operational amplifiers with gm control. IEEE International Midwest Symposium on Circuits and Systems, 2006.
12. Cini U., Toker A. (2017) DVCC based very low-offset current-mode instrumentation amplifier. International Journal of Electronics, 104, https://doi.org/10.1080/00207217.2017.1296592.
13. Nielsen J. H., Bruun E. (2004) A CMOS low-noise instrumentation amplifier using chopper modulation. Analog Integrated Circuits and Signal Processing, 42:65–75.
14. Fan Q., Sebastiano F., Huijsing J. H., Makinwa K.A. (2011) 1.8 μW 60nV/$\sqrt{}$Hz capacitively-coupled chopper instrumentation amplifier in 65 nm CMOS for wireless sensor nodes. IEEE Journal of Solid-State Circuits, 46(7):1534–1543.
15. Bruschi P., Cesta F.D., Piotto M., Simmarano R. (2014) A very compact CMOS instrumentation amplifier with nearly rail-to-rail input common mode range. European Solid State Circuits Conference (ESCIRC), 2014.
16. Zamora-Mejía G., Martínez-Castillo J., Rocha-Pérez J. M., Díaz-Sánchez A. (2016) A current mode instrumentation amplifier based on the flipped voltage follower in 0.50 μm CMOS. Analog Integrated Circuits and Signal Processing,87(3):389–398.

17. Douglas E. L., Lovely D. F., Luke D. M. (2004) A low-voltage current-mode instrumentation amplifier designed in a 0.18-micron CMOS technology. Canadian Conference on Electrical and Computer Engineering, 2004.
18. Safari L., Minaei S., Ferri G., Stornelli V. (2018) A low-voltage low-power instrumentation amplifier based on supply current sensing technique. AEU - International Journal of Electronics and Communications, 91:125–131.
19. Safari L., Minaei S. (2017) A low-voltage low-power resistor-based current mirror and its applications. Journal of Circuits, Systems and Computers, 26(11), https://doi.org/10.1142/S0218126617501808.
20. Eldeeb M. A., Ghallab Y. H., Ismail Y., El-Ghitani H. (2018) A 0.4-V miniature CMOS current mode instrumentation amplifier. IEEE Transactions on Circuits and Systems II: Express Briefs, 65(3): 261–265,

Chapter 9
CMIA for Sensor Applications

9.1 CMIA Application for Piezo-Resistive Sensors

9.1.1 Introduction on Piezo-Resistive Sensors

Piezo-resistive sensor is a particular kind of sensor where its electrical resistance changes as a result of applied forces [1–6]. Piezo-resistive pressure sensors exhibit significant advantages over other pressure-sensor types. They yield high-pressure sensitivity and can be fabricated in standard CMOS processes which offer the possibility of on-chip signal processing [1]. Their principle of operation can be explained by considering the resistance of a rectangular conductor expressed by [1]:

$$R_0 = \rho_0 \frac{l}{wt} \tag{9.1}$$

being ρ_0 the resistivity and l, w and t length, width and thickness of the conductor, respectively. A strain applied to the resistor changes its value as follows [1]:

$$\frac{\Delta R}{R_0} = \frac{\Delta l}{l} - \frac{\Delta w}{w} - \frac{\Delta t}{t} + \frac{\Delta \rho}{\rho_0} \tag{9.2}$$

In Eq. (9.2), the first three terms denote the change in resistor due to dimensional changes which is prevailing for metals, while the last term represents the change in resistivity dominant in semiconductors. The piezo-resistive effect of semiconductor materials can be of several orders of magnitudes larger than the geometrical effect and is present in materials like Ge, Si, etc. Hence, high sensitivity strain gauges can be built using semiconductors that can be locally fabricated in bulk silicon through ion implantation or diffusion, considering that p-type silicon and n-type silicon exhibit positive and negative sensitivity on pressure variation, respectively.

G. Ferri et al., *Current-Mode Instrumentation Amplifiers*, Analog Circuits and Signal Processing, https://doi.org/10.1007/978-3-030-01343-1_9

9.1.2 Circuit Description of CMIA for Piezo-Resistive Sensor Applications

In [7] a signal conditioning circuit for piezo-resistive sensors is presented. The circuit is shown in Fig. 9.1. Its first stage is a CMIA. To have a compact and noise-immune structure, a simple 1 bit A/D converter is directly used in second stage to convert the output of CMIA to a digital signal which can be directly processed. The CMIA structure is based on one CCII$^-$, one CCII$^+$ and on four resistors named R_{1a}, R_{2a}, R_{1b}, and R_{2b}. The outputs of CCIIs are connected to form a current subtraction node. Complementary piezo-resistive sensors are shown by R_P^+ (its resistance increases with a pressure) and R_P^- (its resistance decreases with a pressure) and are supplied by constant current sources I_0. Any change in the value of sensors appears as a proportional voltage at the Y terminal of CCIIs. Therefore, sensors are configured in a kind of mixed-mode bridge configuration because they are excited by a current and produce an output voltage signal.

In [7] two operational floating amplifiers (OFAs) are used to construct the required CCIIs. Figure 9.2 shows the symbol of OFA which is a four-terminal current-mode active building block [7, 8]. Its operation is expressed as [8]:

$$I_o^+ = I_o^- = A_g\left(V_{in+} - V_{in-}\right) \tag{9.3}$$

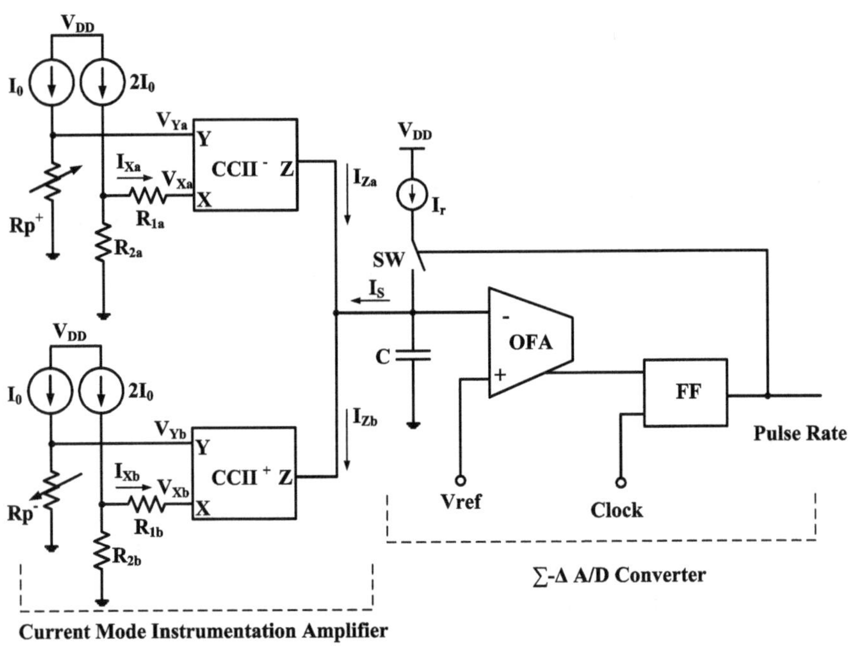

Fig. 9.1 Signal conditioning circuit for piezo-resistive sensor application reported in [7]

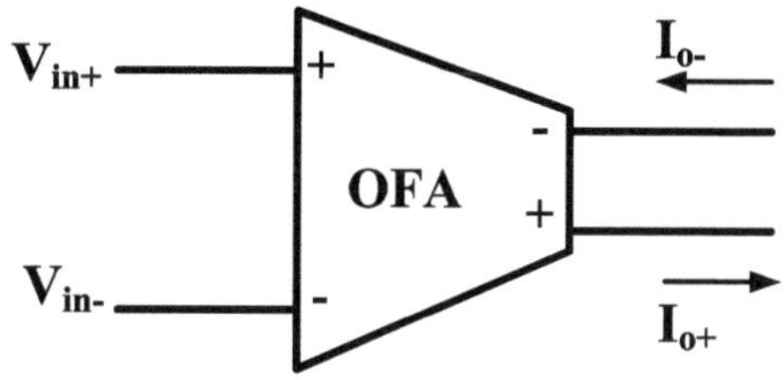

Fig. 9.2 Operational Floating Conveyor symbol [8]

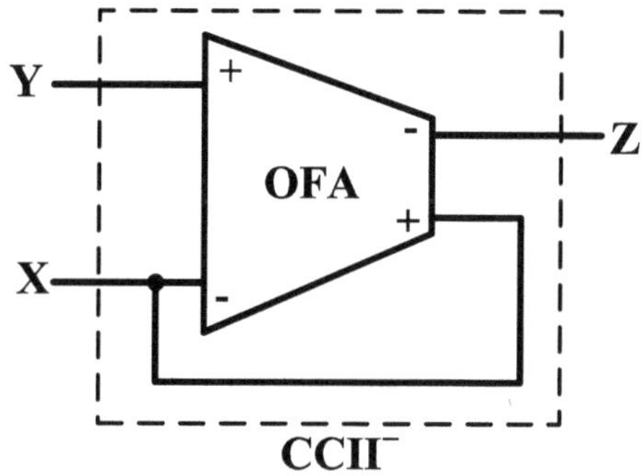

Fig. 9.3 OFA connected in a $CCII^-$ configuration [7]

being A_g the gain of OFA. The OFA exhibits input characteristics similar to Op-Amps; in fact, by assuming a large A_g, the voltage difference between input ports is approximately equal to zero and it can be used in various feedback configurations.

With respect to Op-Amp, an OFA has high impedance current output ports. Figure 9.3 shows the OFA application as a $CCII^-$ reported in [8]. A $CCII^+$ can be easily made by adding extra current mirrors to the $CCII^-$ circuit to invert the output current.

Using the basic equation of CCII ($V_X = V_Y$, $I_Z = \pm I_X$), I_{Xa} and I_{Xb} in the circuit of Fig. 9.1 are found as:

$$I_{Xa} = \frac{I_0}{1+\frac{R_{1a}}{R_{2a}}}\left(2-\frac{R_p^+}{R_{2a}}\right) \tag{9.4}$$

$$I_{Xb} = \frac{I_0}{1+\frac{R_{1b}}{R_{2b}}}\left(2-\frac{R_p^-}{R_{2b}}\right) \tag{9.5}$$

By assuming $R_{1,2a} = R_{1,2b} = R_0$, $R_{p+} = R_0 + \Delta R$, $R_{p-} = R_0 - \Delta R$, from Eqs. (9.4 and 9.5), I_{Xa} and I_{Xb} are expressed as:

$$I_{Xa} = \frac{I_0}{2}\left(1 - \frac{\Delta R}{R_0}\right) \tag{9.6}$$

$$I_{Xb} = \frac{I_0}{2}\left(1 + \frac{\Delta R}{R_0}\right) \tag{9.7}$$

As the result of current subtraction at the output of CMIA, its output current is expressed as:

$$I_S = I_{Xb} - I_{Xa} = I_0 \frac{\Delta R}{R_0} = I_0 SP \tag{9.8}$$

where $SP = \frac{\Delta R}{R_0}$ is called piezo resistor parameter, being S the sensitivity of piezo resistor and P the applied pressure to which the output current of CMIA is proportional.

The second stage is a 1-bit Σ-Δ analog to digital converter made of current source I_r, capacitor C, one OFA and one flip-flop (FF). The comparator formed by OFA operates in positive or negative saturation state. In Fig. 9.1, for zero pressure, I_s will be zero because $I_{Za} = I_{Zb}$. In other cases, the capacitor C is discharged by current I_s or charged by $I_r - I_s$ (assuming $I_r > I_s$). The number of the output pulses (n) in a specified time interval T_{int} is proportional to I_s being:

$$\frac{n}{T_{\text{int}}} = k\frac{I_s}{I_r} \tag{9.9}$$

9.1.3 Error Sources and Compensation Methods for CMIA in Piezo-Resistive Sensor Applications

9.1.3.1 Tracking Errors in Piezo-Resistive Sensors

The tracking errors, due to batch fabrications, cause deviation from the nominal value of piezo-resistors which produces a temperature-dependent offset signal at the output. In the CMIA of Fig. 9.1, resistors $R_{1a,b}$ and $R_{2a,b}$ are used to compensate the tracking errors. As Eqs. (9.4 and 9.5) indicate, if resistors $R_{1a,b}$, $R_{2a,b}$ and sensors $R_P^{\pm}$ have the same deviation, their ratio is independent of errors, so the output current will be free of errors. For this purpose, these resistors are located in the bulk and as close as possible to piezo-resistors. The resistors are also designed to have the same shape and size with piezo-resistors. Therefore with a careful design, the effect of tracking errors is compensated.

9.1.3.2 Errors Caused by CMIA

Non-zero values of input referred offset current (I_{off}) and offset voltage (V_{off}) of the CMIA circuit are other sources of error in the read-out circuit of Fig. 9.1. They produce equivalent offset currents at the outputs of CCIIs expressed in [7] as:

$$I_{offeq1} = I_{off1} + \frac{V_{off1}}{R_{1a} + R_{2a}} \tag{9.10}$$

$$I_{offeq2} = I_{off2} + \frac{V_{off2}}{R_{1b} + R_{2b}} \tag{9.11}$$

where I_{offeq1} and I_{offeq2} are the equivalent offset current at the output of $CCII_1$ and $CCII_2$ respectively. I_{offeqi} and V_{offeqi} for i = 1,2 are the offset current between X and Z terminals and offset voltage between Y and X terminals of the related CCII. Due to the current subtraction at the output of CMIA, these errors appear at the output of CMIA as an equivalent offset current equal to:

$$I_{offeq} = I_{offeq1} - I_{offeq2} \tag{9.12}$$

According to Eq. (9.12), due to the current subtraction operation at the output of CMIA, the effect of offset components of CCIIs is significantly reduced if the two CCIIs are carefully designed to have identical parameters.

9.1.3.3 Errors Caused by Temperature

Piezo-resistive pressure sensors are inherently sensitive to temperature and their output signal and zero pressure offset are varied by temperature. One effective method used in [7] to compensate the temperature effects is related to the use of a bias current source which has a temperature coefficient opposite to that of piezo-resistive sensor. In this manner, according to Eq. (9.8), the output of CMIA is independent from temperature.

9.2 CMIA for Differential Capacitive Sensors

9.2.1 Basic Principle of Differential Capacitive Sensors

Other widely used sensors in industry are the capacitive sensors. In literature [9–12], they have been widely used to detect various parameters such as humidity, pressure, position and displacement, speed, acceleration etc. They are typically formed by two metal parallel plates separated by an insulator layer. The sensor capacitive value can be changed according to a variation of the areas of plates, the gap between two plates or the dielectric constant of capacitor, possibly related to a variation in the chemical, physical or biological quantities that the sensor is able to detect. Among the capacitive sensors, differential or ratiometric sensors are employed to reduce resolution problems related to low capacitive variations and common-mode signals [12]. Figure 9.4 shows the schematic diagram of a differential capacitive sensor which is constituted by a series connection of two capacitors C_1 and C_2. The common node C is used as excitation node and nodes A and B are connected to signal conditioning circuit. The change in the two capacitances is equal and opposite. If the measured affects the distance between the plates, as is shown in Fig. 9.5a, the value of capacitors is expressed as:

$$C_1 = C_0 \frac{1}{1 \mp kx} \quad (9.13)$$

$$C_2 = C_0 \frac{1}{1 \pm kx} \quad (9.14)$$

where C_0, k and x are the baseline value of C_1 and C_2, the sensitivity of the sensor and the displacement, respectively.

If the measurand affects the overlapping area of capacitors plates (Fig. 9.5b), the values of capacitors are linear function of x, as follows:

$$C_1 = C_0 (1 \mp kx) \quad (9.15)$$

$$C_2 = C_0 (1 \pm kx) \quad (9.16)$$

As it is stated in [13], for both the cases, the variable x can be expressed as follows:

$$x = \frac{C_1 - C_2}{C_1 + C_2} \quad (9.17)$$

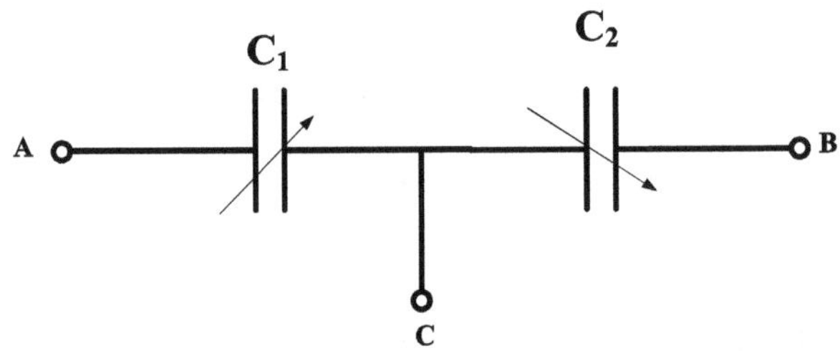

Fig. 9.4 Schematic diagram of differential capacitive sensor [12]

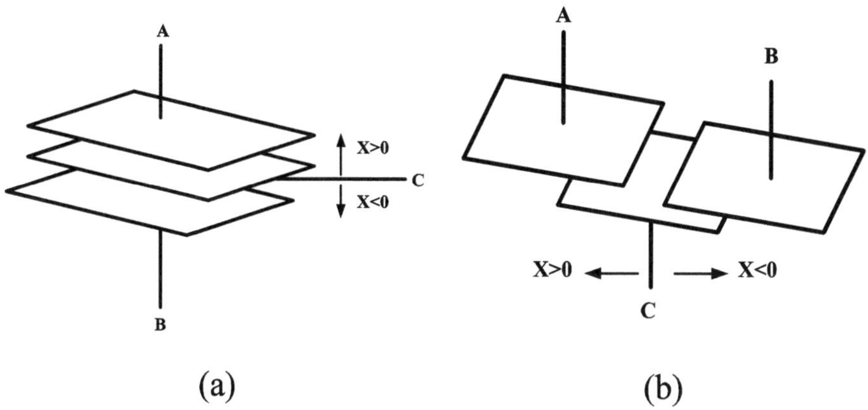

Fig. 9.5 Differential capacitive sensor structure (**a**) when the measured changes the distance between plates; (**b**) when the measurand changes the overlapping area of plates [13]

9.2.2 *CMIA for Signal Conditioning of Differential Capacitive Sensors*

Figure 9.6 shows the CMIA circuit employed in [12] for differential capacitive sensor applications. The CMIA circuit is based on one $CCII^+$ and one $CCII^-$. The Y terminals of CCIIs are connected to ground while their Z nodes are connected together to form a current subtraction node. Capacitors C_1 and C_2 form the differential capacitor sensor. An extra capacitor C_3 is connected in parallel to C_1 with the aim of distinguishing between positive and negative changes in x. The middle port of sensor is connected to a current source and the other ends are connected to the inputs of CMIA (X ports of CCIIs). Therefore, both ends C_1 and C_2 are equipotential and behave as parallel capacitors.

By assuming ideal CCIIs ($V_X = V_Y$, $I_Z = I_X$ for the $CCII^+$ and $I_Z = -I_X$ for the CCII−), a straightforward analysis gives the output voltage as:

$$V_{out} = \left(R_L \parallel R_{z1} \parallel R_{z2}\right) I_{ref} \left[\frac{C_0}{C_0 + C_3} x + \frac{C_3}{C_0 + C_3} \right] \tag{9.18}$$

where R_{z1} and R_{z2} are parasitic impedances at Z ports of $CCII_1$ and $CCII_2$, respectively. As explained in previous section, C_0 and x are the baseline value of C_1 and C_2, and the displacement, respectively.

As Eq. (9.18) shows, the output voltage is proportional to x. This circuit requires a voltage buffer at output for practical use.

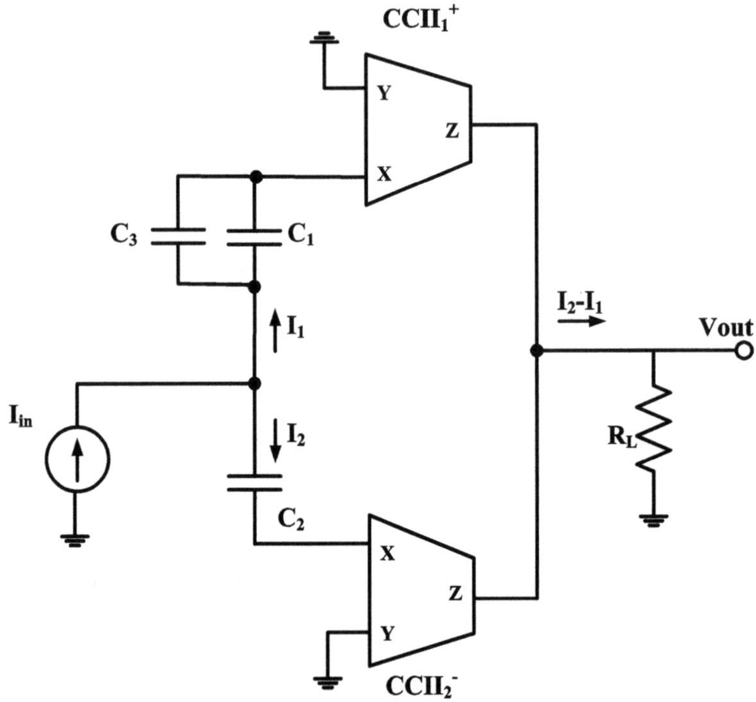

Fig. 9.6 CMIA application for differential capacitive sensor application [12]

9.3 CMIA for Chemical Sensor Applications Based on Ion Sensitive Field Effect Transistor-ISFET

9.3.1 Introduction on Ion Sensitive Field Effect Transistor–ISFET

The ion sensitive field effect transistor (ISFET) is a chemically sensitive field effect transistor, widely used to measure ion concentration in a solution and conventionally referred to as a pH sensor. As discussed in [14–16], it is a small size device and can be readily fabricated using a conventional MOS process. Figure 9.7 shows the cross section of an ISFET. Unlike the conventional MOSFET, instead of a metal gate, it uses an electrolyte solution in which a reference electrode is immersed. Its operation is similar to conventional MOSFET and the theoretical description only requires a minor change compared to a MOSFET. In [16], the threshold voltage (V_{TH}) of ISFET is expressed as:

$$V_{TH} = K_1 - \psi_0 (pH) \tag{9.19}$$

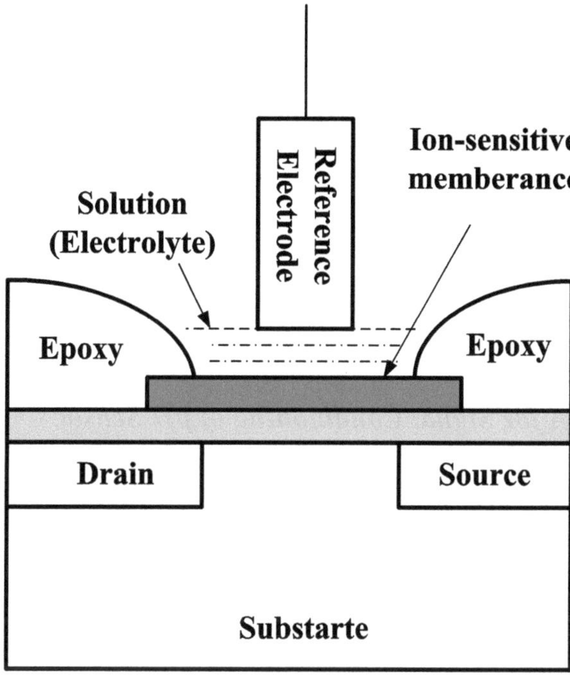

Fig. 9.7 Cross section of an ISFET [14]

where $\psi_0(pH)$ is the pH dependent term and K_1 is a summary of all pH independent parameters. More in detail, the parameter $\psi_0(pH)$ is the potential difference between the insulator surface exposed to the electrolyte and the bulk of the electrolyte itself. The ISFET is commonly used in linear region where $V_{DS} \ll (V_{GS} - V_{TH})$ and the relation between its drain current and nodes voltages is expressed as Eq. (9.20) with usual meaning of symbols.

$$I_{DS} = k\left[\left(V_{GS} - V_{TH}\right) - \frac{V_{DS}}{2}\right]V_{DS} \tag{9.20}$$

where $K = C_{ox}\mu\frac{W}{L}$.

For $\frac{V_{DS}^2}{2} \ll \left(V_{GS} - V_{TH}\right)$, and constant V_{GS}, I_{DS} is expressed as:

$$I_{DS} = K\left(V_{GS} - K_1\right)V_{DS} + V_{DS}\psi_0\left(pH\right) \tag{9.21}$$

In Eq. (9.21), the first term is constant while the second one detects any change in electrolyte pH according to a corresponding change in the drain current I_{DS}.

However, as it is stated in [16], the dependence on measurement temperature and threshold voltage as well as the technological difficulties associated with the packing of miniature reference electrodes limits the commercial viability of ISFET applications. These difficulties can be overcome by using a differential ISFET which consists of an ISFET and a reference FET (REFET) having the same ISFET characteristics but with less sensitivity to the pH variation, both integrated in the same device structure. As an example, an ISFET sensor covered with a Si_4N_3 sensitive layer, has a pH sensitivity of 7.2 μA/pH, while the REFET which is covered by a S_iO_2 layer, has a pH sensitivity of about 1.7 μA/pH [16]. A schematic representation of differential ISFET is shown in Fig. 9.8.

9.3.2 CMIA for Signal Conditioning of pH Sensor

Figure 9.9 shows the Operational Floating Current Conveyor (OFCC)-based CMIA for pH sensor application reported in [16]. The circuit consists of two OFCCs and two resistors. OFCC is a five-terminal current-mode building block, where Y is a high impedance voltage input, X is a low impedance current input, W is a low impedance voltage output and Z terminals are high impedance current output ports. The relationship between current and voltages of terminals are the following:

$$i_y = 0,\ \ v_x = v_y,\ \ i_{z+} = -i_{z-} = i_w \text{ and } v_w = Z_t i_x \tag{9.21}$$

where Z_t is open-loop transimpedance gain to produce an output voltage at terminal W. Ideally Z_t is infinite and correspondingly, input current at terminal X is zero.

The inputs of CMIA are connected to the ISFET sensor and the REFET sensor. The currents passing through ISFET and REFET are conveyed to the output of

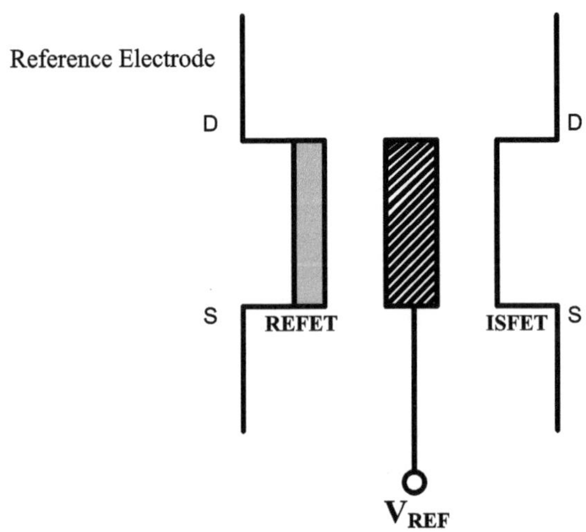

Fig. 9.8 Schematic diagram of differential ISFET [16]

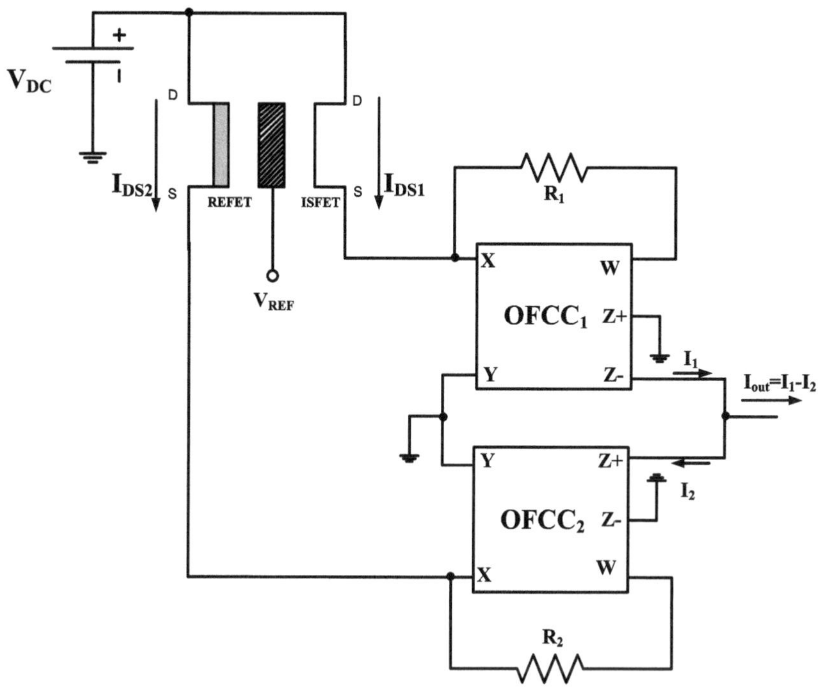

Fig. 9.9 OFCC based CMIA for differential ISFET application [16]

CMIA by $OFCC_1$ and $OFCC_2$, respectively, where they are subtracted. Therefore, for $I_{DS1} = I_{DS2}$, the current difference at output node is negligible resulting in a high common-mode rejection ratio. In this configuration, the source terminals of sensor are fixed at ground because of the voltage tracking between X and Y terminals of the OFCC. The drain voltage is also set at a constant voltage V_{DC}. This forces the drain-source voltage of sensor to remain at constant value as well as the gate voltage of sensor which is fixed at V_{REF}. Therefore, the operating point of sensor is appropriately set. The threshold voltage and output current of ISFET (I_{DS1}) varies as a result of any change in the pH of the solution. As it was mentioned, ISFET is more sensitive to pH variation compared to REFET, therefore, the variation in the threshold voltage and output current of REFET (I_{DS2}) are much smaller. I_{DS1} and I_{DS2} are conveyed to the output terminals of OFCCs where they are subtracted resulting in output current of:

$$I_{out} = I_1 - I_2 = \alpha_1 I_{DS1} - \gamma_2 I_{DS2} \tag{9.22}$$

where α_1 and γ_2 denote the non-unity current gains between X, Z^+ and Z^- terminals of corresponding OFCC, respectively.

9.4 Chopper Stabilized CMIA Designed for High Precision Temperature Sensor Application

A low noise and low offset CMIA for high precision temperature sensor application is presented in [17]. The low value of detectable signal needs a very low noise CMIA. As the bandwidth in this application is very low, the nested chopping technique is utilized in [17] to eliminate the 1/f noise and suppress the offset. The circuit consists of input modulators, a CMIA made of two CCIIs and output modulators. The modulators are implemented by a set of switches. The CMIA circuit is shown in Fig. 9.10. The input voltage signal at Y terminals of CCIIs is conveyed to X terminals where is converted to current by R_i. The current produced across R_i is transferred to Z terminals and converted to voltage again by R_o producing the gain of:

$$A_v = \frac{V_{in+} - V_{in-}}{V_{out+} - V_{out-}} = \frac{R_o}{R_i} \tag{9.23}$$

The input signal is modulated twice by two square wave signals mixing it to the odd harmonics of the chopping frequencies. The choppers are implemented by a set of transistor-made switches. In [17] the high chopping frequency is selected at noise corner frequency.

As it is discussed in [17] two main sources of offset in chopper amplifier are charge injection spikes from choppers and non-idealities in chopping waveforms. Minimizing lower chopper frequency and the dimensions of the chopper transistors result in better offset performance. For low flicker noise, the CCII input transistors are chosen of PMOS type and their size is increased as much as possible; however, by increasing the size of transistors, the input capacitance is increased by the same amount leading to a higher offset. Choosing low chopping frequencies is helpful in minimizing the non-idealities in chopping waveforms (i.e., average value of the mismatch or skew in one period). It is worth noting that, as the input choppers receive the low value input signal, they are the most noise critical part of the total instrumentation amplifiers. To this aim, the output choppers receive the amplified signal making less noise critical part of the circuit.

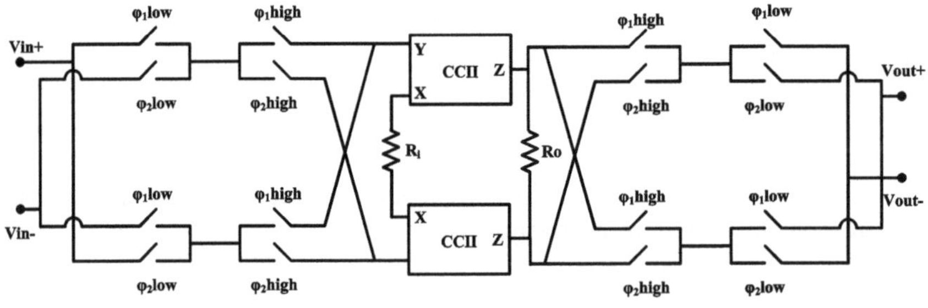

Fig. 9.10 High precision CMIA of [17] based on nested chopping technique

References

1. Kanda Y. (1991) Piezoresistance effect of silicon. Sensors and Actuators A, 28: 83–91.
2. Suhling J. C., Jaeger R. C. (2001)Silicon piezo-resistive stress sensors and their application in electronic packaging. IEEE Sensors Journal, 1(1):14–30.
3. Hammerschmidt D., Schnatz F. V., Brockherde W., Hosticka B. J., Obermeier E. (1993) A CMOS piezo-resistive pressure sensor with on-chip programming and calibration. IEEE International Solid-State Circuits Conference Digest of Technical Papers, 128–129.
4. Lee D.-W., Choi Y.-S. (2008) A novel pressure sensor with a PDMS diaphragm. Microelectronic Engineering, vol. 85, pp. 1054.
5. Akbar M., Shanbaltt M. (1992) Temperature compensation of piezoresistive pressure sensors. Sensor and Actuators A, 33(3):155–162.
6. Matsuno M., Adachi S., Nakayama M., Watanabe K. (1993) A bridge circuit for temperature drift cancellation. IEEE Transactions on Instrumentation and Measurement, 42(4):870–872.
7. Vlassis S., Siskos S., Laopoulos T. (1999) A piezoresistive pressure sensor interfacing circuit. IEEE Instrumentation and Measurement Technology Conference, 1:303–308.
8. Laopoulos T., Siskos S., Bafleur M., Givelin P., Tournier E. (1995) Design and applications of an easily integrable CMOS operational floating amplifier for the MegaHertz range. Analog Integrated Circuits and Signal Processing, 7:103–111.
9. Chatterjee K., Mahato S. N., Chattopadhyay S., Daxler D. (2014) Capacitive high accuracy mass measuring system using capacitive sensor. Instruments and Experimental Techniques, 57(5):627–630.
10. Toth T. N., Meijer G. C. M. (1992) Low-cost smart capacitive sensor. IEEE Transaction on Instrumentation and Measurements, 41(6):1041–1044.
11. Baxter K. (1997) Capacitive sensors: design and applications. The Institute of Electrical and Electronics Engineers, 1–46.
12. Ferri G., Parente F. R., Stornelli V., Barile G., Pennazza G., Santonico M. (2016) CCII-based linear ratiometric capacitive sensing by analog read-out circuits. Lecture Notes in Electrical Engineering.
13. Barile G., Ferri G., Parente F. R., Stornelli V., Depari A., Flammini A., Sisinni E. (2017) Linear integrated interface for automatic differential capacitive sensing. Proceedings of Eurosensors, 2017.
14. Nobpakoon T., Pijitrojana W., Poyai A. (2013) A new method for current differential ISFET/REFET readout circuit. International Journal of Information and Electronics Engineering, 3(2):141–143.
15. Ghallab Y. H., Badawy W., Kaler K. V. I. S. (2003) A novel pH sensor using differential ISFET current mode read-out circuit. International Conference on MEMS, NANO and Smart Systems, 2003.
16. Ghallab Y. H., Badawy W.(2004) A new differential pH sensor current mode read-out circuit using two operational floating current conveyor.IEEE International Workshop on Biomedical Circuits and Systems, 2004.
17. Voulkidou A., Siskos S., Laopoulos T. (2013) Analysis and design of a chopped current mode instrumentation amplifier. International Journal of Microelectronics and Computer Science, 4(1):6–11.

Index

G. Ferri et al., *Current-Mode Instrumentation Amplifiers*, Analog Circuits and Signal Processing, https://doi.org/10.1007/978-3-030-01343-1

W

MIX
Papier aus verantwortungsvollen Quellen
Paper from responsible sources
FSC® C105338

If you have any concerns about our products,
you can contact us on
ProductSafety@springernature.com

In case Publisher is established outside the EU,
the EU authorized representative is:
Springer Nature Customer Service Center GmbH
Europaplatz 3, 69115 Heidelberg, Germany

Printed by Libri Plureos GmbH
in Hamburg, Germany